Date Due

MAR 3 '99			
APR 12 2000			
APR 6 2001			
MAY 0 6 2004			

DISCARDED

BRODART, CO. Cat. No. 23-233-003 Printed in U.S.A.

PRESCRIPTION FOR THE FUTURE

How the Technology Revolution Is Changing the Pulse of Global Health Care

PRESCRIPTION FOR THE FUTURE

How the Technology Revolution Is Changing the Pulse of Global Health Care

Gwendolyn B. Moore

David A. Rey

John D. Rollins

ANDERSEN CONSULTING

LIBRARY
Pennsylvania College
of Technology

One College Avenue
Williamsport, PA 17701-5799

Knowledge Exchange, LLC, 1299 Ocean Avenue
Santa Monica, California 90401

Copyright © 1996 by Andersen Consulting.
All rights reserved

No part of this publication may be reproduced, stored in a retrieval system, or otherwise transmitted, in any form or by any means, electronic, mechanical, photocopying, recording, or otherwise, without the prior written permission of the publisher. Printed in the United States of America. Published simultaneously in Canada.

Jacket design by Denise Palma, Marsteller Advertising
Text design by Carolyn Wendt

1 2 3 4 5 6 7 8 9-VA-99 98 97 96
First printing June 1996

Knowledge Exchange books are available at special discounts for bulk purchases by corporations, institutions, and other organizations. For more information, please fax Knowledge Exchange, LLC, at (310) 394-7637.

Knowledge Exchange and the pyramid logo
are trademarks of Knowledge Exchange, LLC.

Contents

To H.H.D.L.,
For standing as a symbol for all that is possible
for individuals and for the world.

G.B.M.

To Skip Battle,
My partner, mentor, and friend, who was
responsible for my career focus in health care.

D.A.R.

To Anne, David, Margaret, and Theresa,
For allowing my work to peacefully exist with
our family.

J.D.R.

ACKNOWLEDGMENTS

The writing of this book—enriched by the insight and experience of many to whom we owe a debt of gratitude—has been a truly collaborative effort. The clients of Andersen Consulting lead the list of those who contributed. By sharing their challenges and victories with us, they have helped us learn what really works and what doesn't.

Dozens of our Andersen Consulting colleagues served as invaluable sounding boards, and all three of us benefited tremendously from their expertise, which represents a broad spectrum of industry, technology, and strategy perspectives. Their contributions ranged from helping to shape and refine our message to reviewing the final manuscript. We extend our thanks to Jim Anderson, Thomas Baubin, Ken Beecham, Christie Blish, Tim Breaux, Jim Brennan, Tom Cobb, Nick Edwards, Peter Fuchs, Richard Griffiths, Jim Hudak, Bruce Johnson, Phyllis Kennedy, Ann Klein, Mark Laleike, Sandi Marks, Jim Murphy, Terry Neill, Michael Palmer, Carla Paonessa, Jim Politoski, Rudy Puryear, William Reinfeld, Brian Reynolds, Steve Rushing, Alan Salter, Iain Somerville, Allen Stiles, Joe Villa, Diane Wilson, and Dr. John Zelcer.

We recognize, in particular, Brian Pappas, who helped shape our thinking about how information technology enables business strategy; Bill Higbie and Wayne Piez whose day-in-the-life scenario inspired Chapter 6; Charles Roussel, for his ideas about the organizational impact of technology; and Dr. Michael Eliastam, whose extra effort to our manuscript made sure that our medical and scientific references were correct.

From outside Andersen Consulting, we extend very special thanks to James Bradley, President of Health Systems Integration, Inc., and Max Hopper, Principal and CEO, Max D. Hopper Associates, for their review of the manuscript and insightful comments. Also, four individuals from Astra Merck Inc. earn thanks for sharing their experience with us: Alana Edelmann, Bob Holmes, Ken Murtha, and Nancy Wysenski. Helen Rees, our literary agent, helped us turn our idea for a book into reality. And we thank Michael Dennis and Nancy Garuba for their sound advice.

Finally, we dare say that the book would never have reached the printing press without the day-to-day support of these colleagues: Leslie Heidenreich, Todd Israel, and Jan Boscacci, who were especially instrumental. We're also grateful to Julie Husband, Tom Johnson, Stacey Jones, Maryanne Keeney, Sara Nuernberg, Nancy Royce, and Marya Witt. Thanks also to Donna Carpenter, Christina Braun, Maurice Coyle, Elyse Friedman, Erik Hansen, Martha Lawler, Stephen McCauley, and G. Patton Wright for their talent and patience. Without their efforts we would still be writing!

Yet, in spite of all of that help, we, the authors, assume responsibility for the book's messages and the value they may bring to the reader.

G.B.M.
D.A.R.
J.D.R.

— FOREWORD —

Even as I write this, stupendous changes are occurring in the health care industry. This morning's paper greets me, for example, with the news that Aetna Life and Casualty will acquire U.S. Healthcare for $8.9 billion. And just recently, Columbia/HCA announced that it plans to merge with Blue Cross/Blue Shield of Ohio. In business deals that rival some of the mammoth consolidations in the banking, entertainment, and telecommunications industries, health care organizations are joining forces and dislodging many of our long-held and most unshakable business assumptions.

Many forces—medical, economic, and social—motivate those mergers, acquisitions, and partnerships, but they wouldn't be possible were it not for the simultaneous revolution in information technology. Over the past few decades, technological innovations have transformed our cultures, our business practices, and our world views. Information technology, however, is not the be-all and end-all of these changes. It's merely the chief means of getting from where we are to where we need to be: changing the way health care is conceived and delivered.

In *Prescription for the Future,* Gwen Moore, Dave Rey, and John Rollins probe deep beneath the foundations of the health care industry. Their electrodes have recorded a Richter-scale measurement of the tremors and aftershocks of information technology. Although the authors wrote the book primarily for people involved in health care—industry managers, physicians, nurses, insurers, suppliers, and the like—those who work in other industries that feel the ripples of information technology will find *Prescription* equally relevant to their situations.

Nowadays, that's just about *everybody.* Throughout my career, I have had many opportunities to watch, first-hand, the development of technology as a force for industrial change. Moore, Rey, and Rollins name these forces. They describe how technology gives us the agility, speed, reach, and insight we need to be competitive. The next generations will take those business assets to new heights; higher than anything we can imagine at this moment. While no one can predict the future with absolute certainty, of course, I feel confident to assert that the next several decades will see astronomical advances in our ability to maximize the power of data and knowledge: We will discover, learn, produce, and utilize more than ever before.

Distinguished from other books on this topic, *Prescription for the Future* presents a holistic vision of the requisite elements of change to create a future state that the authors have dubbed the Infocosm. The Infocosm has vast potential to alter our lives and the way we do business. Yes, there is an unnerving sense of disruption here, and it threatens our familiar lives. I am reminded, however, that, in Chinese, one word means both "threat" and "opportunity." While nothing that is worth pursuing is risk-free, not engaging in the difficult pursuit is to put one's business at great peril.

If I were to advise a group of CEOs about the benefits of business innovations that information technology makes possible,

I would tell them this: Technology is neither a quick fix nor a short-term investment that will make them success stories overnight. Rather, this revolution in information technology requires one's long-term commitment: It encompasses the deepest changes in the way the world works. We face a cultural upheaval of the first order. Those of us lucky enough to have been involved in the beginning of this change have been picking only the lowest-hanging fruits. Those who stand on our shoulders extend our grasp, just as the health care professionals who adopt the authors' holistic view of transformational change will enable their organizations to find creative solutions to problems both past and present. In contrast to the short-sighted who choose not to investigate the enabling features of this technology, the more daring leaders' organizations will be more competitive, more innovative, and more responsive to progress. "In the Infocosm or Out of Business" reads one chapter title in this book. No phrase better captures the essence of what information technology means for industry leaders who want to win.

Max D. Hopper
Dallas, Texas
April, 1996

INTRODUCTION

Information technology is driving enterprises down a road whose end is not clearly in sight. All around us lies unfamiliar territory, and some of us understandably feel anxious about driving into it. Savvy travelers, however, are speeding down this road where they see it merging gracefully with the information superhighway.

There is both good news and bad news in the emergence of technology as a force for change. The good news is that new technologies are smashing the barriers of time, place, and form, allowing organizations to create new business processes that extend beyond the confines of their own enterprise. From this "creative destruction," as the great economist Joseph Schumpeter would call it, comes wholly new kinds of corporate assets. One of these new assets is insight—the ability to access, filter, analyze, and synthesize diverse information and experience to allow us to see new patterns, new relationships, new ideas.

But these changes bring our old assumptions about the rules of business crashing down around us. Now, to survive, leaders and entrepreneurs must either come to grips with all that is possible in a new landscape or watch from the sidelines as their competitors rush on to victory.

Why is this happening now? Because it *can* happen through information technology. And with the possibility is the challenge to *do* something. Both the cause and the solution are rooted in technology.

This book is not about technology *per se,* but about the effect it is having in shaping business imperatives. It is a book that has grown from the lessons we have learned from helping our clients come to grips with these challenges by trying to separate the good from the bad, the environmental realities from individual choices. We see technology as both *destabilizer*—by challenging old assumptions, for example—and *enabler,* giving rise to new capabilities and opportunities.

This book is focused through the lens of the health care industry, but it is not another book about what ails health care. We have chosen to illustrate this destabilizing/enabling paradox in health care because this industry is particularly challenged with pressure to reduce costs, improve quality, and redefine how processes are executed both within and across organizations. Yet, this book does not attempt to grapple with the complex social, political, and ethical questions so prominent in industry debates today.

By sharing our experiences, we hope to help health care professionals and policy makers to see what we see: the emergence of the Health Care Infocosm. The Infocosm is neither a thing nor a place, but the future of information technology coupled with changes in business structures and strategies, processes, and people. All these come together to create entirely new ways to compete, to do business, and to deliver wellness and care services. The Infocosm will require individual companies around the globe to operate differently in the new competitive space that the technology will create.

The full potential of the Infocosm is still years away, perhaps not completely operational until the first decade of the twenty-first century. Nevertheless, all organizations must begin preparing now for the sophisticated, technology-enabled environment that the Infocosm will represent. This information technology—and all that it makes possible—is no longer something "nice to have." Rather, it is a necessity for survival.

We have written this book for everyone who must come to terms with this new territory: health care executives and policy makers, health care managers, physicians and other care providers, and, equally important, consumers and patients themselves. But this book is not aimed only at the health care industry, for all that is true in health care is true in every other industry where seismic waves of technology are shaking long-held assumptions about industries and the roles of companies within them. And the lessons are equally true around the world. As a colleague of ours from Taiwan reflects, "The challenges here are the same as those in other markets and industry sectors around the world. As Taiwan becomes a developed and open economy, customers can afford more and are demanding more. In health care. Everywhere."[1]

What does it take to navigate successfully in the Infocosm? Answering that question is the aim of this book.

The first chapter expands on our paradox. We'll show how technological innovation is driving business change. And as this revolution speeds ahead, we'll see the beginnings of the Health Care Infocosm. In this new world, our familiar business context will give way to a radically different environment where care is delivered and business is conducted according to a new set of rules.

In Chapter 2, we describe the journeys by which enterprises are transformed. These journeys must simultaneously define and execute changes in strategy, processes, people, and technology needed to make the Infocosm a reality for each and every company. We illustrate the interdependence of these elements by observing the transformational change undertaken by a representative industry organization that we call El Dorado Health Care.

In the third chapter, we show how the rapid changes in information technology, which are defining the Infocosm, make the strategy-creation context for executives increasingly unstable. Our strategy formulation framework is illustrated by an executive team from an integrated health care delivery system that must come to grips with setting their own strategic direction. The framework defines an iterative and cascading process that is actively informed by an understanding of the new capabilities and possibilities created by information technology. Our approach to strategy-setting explicitly incorporates analyses of the business landscape, a challenging of long-held assumptions, setting strategic intent, and charting a course of action.

In the fourth chapter, we discuss the critical importance of conceiving of business processes that mirror an organization's vision. After all, it is new visions, enabled by new technologies, that come together within business processes to create new organizational assets and capabilities. We describe the critical need to extend core competencies through "plug-and-play" cross-enterprise processes that are complementary and mutually beneficial and to imbue these processes with information technology.

The impact of all of these changes on people and organizations is the subject of Chapter 5. Traditional relationships are giving way to ever-changing ones characterized by cooperation and collaboration. These new relationships require everyone to learn to work and manage differently. This chapter grapples with new ways of valuing knowledge and with how organizations will need to invent new ways to "hold themselves together" around a compelling vision as they draw important capabilities from individuals, groups, and organizations with whom they are affiliated but don't "control." Finally, we show how both employers and employees will need to reskill in the age of the Infocosm. They recognize that smart labor is a competitive advantage for both and that failure to provide a learning environment will put them at risk.

In Chapter 6, we jump to the future. We follow Joe McAlister, a fictional football player, through his own personal health care journey. In this excursion into the year 2001, we see how the Health Care Infocosm shapes his experiences and how Joe is but one of its many beneficiaries.

In Chapter 7, we summarize the key challenges to survival in the Health Care Infocosm. Here we discuss the factors that are key to the enterprise's success in the midst of this destabilizing/enabling paradox. As difficult as it is to conceive and bring about the type of enterprise transformation we describe here, it is essential for future survival. Before too long, health care companies will either be in the Infocosm, or they will be out of business.

CHAPTER 1

SCIENCE NONFICTION

Having lived her entire life on an uninhabited island, Miranda in Shakespeare's *The Tempest* is awestruck by her first encounter with human beings: "Oh brave new world that has such people in it!" she exclaims. (V.i.183–84)

Those of us observing the beginnings of the revolution in the health care industry in these last years of the twentieth century can easily comprehend her untempered joy, excitement, and enthusiasm. Only yesterday, what seemed the stuff of science fiction fantasy is suddenly reality. Last month's remote possibilities are this month's givens. Hollywood's special effects in movies only a few years old seem clumsy in comparison to today's computer-generated virtual reality. Similarly, yesterday's technology is far surpassed by information technology now in everyday use in large medical centers and small health clinics from Boston to Brisbane, Tokyo to Tel Aviv, Rome to Rio. And this is only the beginning.

There are breakthroughs all around the world. In the time it takes you to read this chapter, dozens of lives will have been saved by the application of information technology to health care. A doctor in rural Africa will access updated treatment protocols for a suffering AIDS patient. A health clinic worker in suburban Maryland will present a teen mother with a printout of the latest nutritional data from the Harvard School of Public Health. An oncologist in São Paulo will watch real-time pictures of a colonoscopy taking place in a small town thousands of miles away.

But even the most optimistic and forward-looking among us must occasionally stop to wonder where this revolution is taking us. Shakespeare's Prospero, whose life experiences tainted his vision, reminds Miranda that only the naïve view a new world with unbridled enthusiasm. Every new opportunity brings new responsibility. Change inevitably brings challenge. Each possibility harbors a potential pitfall. How can we embrace the brave new world without being suffocated by it? How can we follow the lead of information technology without going astray? How can we step boldly and confidently into tomorrow?

The future belongs to the prescient, the prepared, the informed. With global technology advancing with every blink of the eye, only the most knowledgeable and well-primed will survive. Stragglers will be left behind; the incoming tide will overcome the hesitant. Our purpose is to guide leaders and entrepreneurs into tomorrow with today's best and most up-to-the-second information. Our goal is to help assure your survival, to help clinch your success. Leave your biases behind. Check yesterday's assumptions at the door. The brave new world we are about to enter demands a fresh perspective and an innovative way of thinking about time and space.

Brave New Words for Brave New Worlds

The health care industry is making extraordinary use of remarkable advances in information technology and simultaneously redefining itself. Physicians, nurses, administrators, insurance agents, pharmaceutical executives, suppliers, and health care consumers are all being asked to prepare for the exciting possibilities in the delivery and management of health care.

The results of those advances and that redefinition will be what we call the Health Care Infocosm. The full potential of the Infocosm will unfold in the months and years to come. Nevertheless, the health care industry, already caught in the flood of information and technologies, must already begin to find better ways to manage and direct those resources into channels that will lead to innovative solutions.

Neither a thing nor a place, the Health Care Infocosm is the new information technology–enabled environment of the future. The Health Care Infocosm will link members and caregivers to one another and to critical preventative and medical information—at any time, in any place—as the following vignette suggests:

JORDAN AND THE MAYO CLINIC

In Jordan, kings and citizens alike can receive medical care from physicians at the Mayo Clinic in Minnesota. King Hussein has his checkups through a satellite link that bounces images of blood vessels, chest x-rays, and ultrasound records, while he and his Mayo-based doctor discuss how he has been feeling. Jordanian physicians will also be educated by satellite from their Mayo colleagues on breast cancer, heart rhythm, and malignancies. The Mayo Clinic now plans other international links to Greece and possibly Argentina, Chile, and Colombia.[1]

> In a rural area at night a child is awakened by an asthma attack. Her mother dials the interactive telecommunications connection for medical care support, and she describes the child's condition. A nurse at the regional medical center 100 miles away instructs the mother to attach telecom probes to the child to measure the child's temperature, blood pressure, and pulse. The nurse listens through an electronic stethoscope to the child's breathing and examines the patient through a high-resolution viewer.[2]

This vignette is no longer far-fetched thinking. There is no doubt that both the increased power of the microprocessor and its falling prices over the last fifteen years have been key drivers of change. But there are emerging drivers, which promise even more dramatic change.

First, the computing industry is abuzz with talk of bandwidth, the term used to measure the size and power of a communications connection. We can see that the cost of bandwidth will decline even faster than the cost of computing.

Second, there are the emerging object technologies, which can successfully address the coordination, classification, and distribution of widely distributed data. These will allow a degree of shared access to information that will revolutionize interactions among companies and individuals. Object technologies will also enable information systems to be much more independent of today's business processes, allowing a new level of flexibility in supporting process innovation.

These technological drivers will form the foundation of the Health Care Infocosm. Upon that foundation, today's disparate organizations will come together to develop new capabilities and competencies, products and services, markets and relationships, and processes and structures. Moreover, organizational walls will become increasingly transparent as cross-organization processes emerge and flourish.

Already, we can see the impact of processing power, bandwidth, and global communications in the explosive growth of the Internet. The Internet is an excellent early predictor of how global infrastructure ties people, companies, and industries together. There are as many uses of the Internet—the forerunner of an international information infrastructure—as there are people who access it. The Health Care Infocosm will evolve from today's Internet, but with its own particular emphasis. The health care industry will require a unique information infrastructure to enable interorganizational processes and on-demand collaboration. Only with that infrastructure in place will the health care industry realize its currently untapped potential for delivering higher quality health care and wellness in a more cost-effective way.

The emergence of the Health Care Infocosm will change the way we do business by making communication and collaboration available any time, any place, anywhere in the world. Eventually, health care functions and processes will occur across and among business entities to a far greater degree than they do today. We dream of an information infrastructure and a health care services marketplace that allows the best clinical and business capabilities, regardless of where they exist, to be seamlessly plugged together to serve consumers. But before we achieve this "plug-and-play" capability, the industry must undergo some standardization. The requisite standards to achieve that care-delivery ideal will evolve differently from place to place. These will coalesce as *de facto* standards, as we see happening today with many aspects of the Internet. In any case, we will build the Health Care Infocosm from the twin foundations of global connectivity and emerging object technology standards.

Voltaire lived in the Age of Enlightenment; Bach composed in music's Baroque period; Botticelli painted during the Italian Renaissance. Now, in the late twentieth century, we are all standing at the threshold of the Age of the Infocosm. Information technology—coupled with changes in business strategies, processes, and people—is creating entirely new ways to compete, to do business, to deliver health care, and to serve society. For health care, Infocosm refers not only to the technology but also to this emerging, transformed business landscape and health care culture.

To bring such scenarios to life, the Health Care Infocosm will embody a number of capabilities and characteristics:

- In the Health Care Infocosm, security of information will be assured although multiple authorized users have access to essential data. Perhaps the single most significant use of information technology will be in the creation of a ubiquitous electronic medical record for every "citizen-patient," to borrow a term from Dora B. Weiner's study of health care in Paris during and after the French Revolution.[3] Such a record will provide a life-long assessment of an individual's health—illnesses, treatments, allergies, surgeries, and programs for maintaining wellness.[4–6] The ubiquitous nature of such records will form the basis for the new, interorganizational care delivery processes that will dramatically reduce costs and improve outcomes in the industry.

- In the Health Care Infocosm, we will ultimately be able to aggregate clinical information, particularly critical for the purposes of disease management and outcomes research. At present, our information systems are severely limited both by the technology itself and by the sensitivity of the information the systems contain. If the Health Care Infocosm is to be successful, we must not only enable the researcher to collect data through blind access to electronic medical records, but also allow multiple caregivers simultaneously to retrieve and record information as needed.

- In the Health Care Infocosm, no single organization will have an advantage over any other, for example, by constructing a network that only some participants can use. In fact, we don't believe that such restrictive efforts can succeed. As in the early days of computer technology, competitors will try to gain near-term advantage by constructing proprietary elements of the Health Care Infocosm. There will be inevitable competition to "own the customer" or "control the community." While those early efforts will help to pave the way, they will surrender to common, standards-based solutions that welcome all companies and deliver on the promise of quality, affordable health care. The Health Care Infocosm will be nonproprietary in terms of both participating organizations and the technologies involved.

- In the Health Care Infocosm, highly effective collaboration among clinicians and health care organizations will be commonplace. That capability will force health care organizations to treat data, information, and knowledge in new ways. Rather than remaining isolated, health care professionals will collaborate to create mutually beneficial insights from their respective stores of information. Where information was previously departmentalized or compartmentalized, in the future it will be available both inside organizations and across them.

These capabilities have long been wanting in the health care industry. In the Health Care Infocosm, we will have them at last.

Information technology is doing more than redefining the industry, however. It's also destabilizing it, subjecting today's health care organizations to as many new risks as potential rewards. The key health care players are repositioning themselves in anticipation of the changes this revolution brings. The rules by which they play are being rewritten even as the players formulate their strategies. The roads for improved access to medical care are being paved, but it's not entirely clear who knows which ramps to take. Every organization in the health care industry must come to grips with those changes. Now is not the time to stand still.

Much is required, they say, of those to whom much has been given. The Infocosm will give us much—and make proportionally sweeping demands. The kinds of information technologies we're describing will not be merely desirable: they will be essential. To put it bluntly, it will be impossible to compete in the Age of the Infocosm without embracing the changes it will demand. Nothing seems more threatening than change, and the bigger the change, the more ominous the threat sounds. The aim of this book is to help you understand the complexity of those challenges and to minimize apparent menaces.

One challenge calls for the creation of order out of the chaos that characterizes the present world of technology. Indeed, our modern word "cosmos" is derived from the Greek *kosmos,* the core meaning of which was "order." This is what we strive for in the Infocosm: order. Mere data without the technology to use it well is fragmented and inaccessible. But when technology transforms data into information, it facilitates informed decision-making. Shared and easily accessible at any point of need, the information assumes still greater value, helping us to provide better business, better health care delivery, and better life for all.

Yet, for a time, as the Health Care Infocosm evolves, we're likely to see more disorder than order. The Health Care Infocosm will dissolve the barriers of time, place, and form, and in so doing will reconfigure the traditional roles of health care professionals. So let's take a closer look at what will happen when those traditional barriers fall.

Time and the Infocosm

Technology, in all of its many guises, has already gone a long way toward dissolving the once familiar structure and constraints of time. Thanks to ATM cards, the term "banker's hours" is practically obsolete worldwide: We can withdraw, deposit, and transfer funds at any hour of the day or night in any country around the world. Sophisticated telephone systems allow us to do business on the phone twenty-four hours a day, 365 days a year. Internet and E-mail connections slice through time zones and date lines.

In the health care industry, dissolving the barriers of time will put critical information at the fingertips of care providers exactly when they need it. In a medical crisis, for example, a team of paramedics can use information technology to monitor the physical condition of a patient in a coma, review his or her vital statistics, and search all cases that might have a bearing on that individual's condition—all in the time it takes for the doctor to scrub. Care providers will be able to review a person's complete medical record, access information about wellness activities, determine what procedures have been performed in the past, quickly review current or past medications, and ascertain what allergies he or she may have. No more delays while caregivers haul charts out of filing cabinets and contact numerous sources to consolidate data. The experience and quality of care improve simultaneously. The medical staff is in better control and can save precious seconds.

Outside of an emergency situation, equally impressive results will come as time barriers fall. Jack Sandman, a longtime health care executive, describes the interaction between a physician and a patient in a hospital as a twenty-four-hour cycle:

> The doctor visits once a day, makes a diagnosis, gives orders, expects them to be carried out. Twenty-four hours later, he or she

> comes back, and the process is repeated. This continues until the patient is cured or deemed well enough to go home. It's a horrible waste of time. From the afternoon when the doctor's orders are carried out until the next morning when he or she returns, nothing new happens. If we could monitor the patient and transmit results back and forth to the doctor without the maddening delays inevitable with paperwork and related details, we could utilize those wasted hours and compress treatment time.

In the Infocosm, we will.

"Saving time," Sandman points out, "saves money. But it can also save lives. Morbidity and mortality go down as timeliness goes up. Breaking down time barriers is a way to get both cost savings and quality improvement."[7]

We are years from fully realizing the potential of information technology for dissolving time. But many barriers have already fallen. In those and other instances, the promise of the future is already here.

In addition, we illustrate one of the more common freedoms from the barrier of time: being able to "work" whenever it is convenient. In writing this book, the three of us worked independent of time constraints: John, in London, worked early mornings, while in Boston, Gwen slept. When John went home, Gwen worked. Dave, in San Francisco, continued the round-the-clock process. Our work on this manuscript followed the sun around the globe.

Place and the Infocosm

Just as information technology removes the barrier of time, it also redefines the place where health care "happens." Make no assumptions about where tomorrow's population will be educated, examined, or treated; where surgical supplies will be purchased or delivered; where health care professionals will be most needed.

In the Appalachian Mountains of North Carolina, the Watauga Medical Center's videoconferencing permits timely oncology consultations with specialists in major metropolitan areas. In this case, removing the barrier of place means moving health care to the patient or the specialist. It means using telemedicine to monitor wellness activities and to treat illness. Building on computer and communications technology, telemedicine has already altered the way health care is provided—especially in remote areas. It has transformed the way the services of specialists are used, and

REMOTE EXAMS WITH FEELING

From 300 miles away, Dr. Gary Doolittle chats with a patient and checks her lungs and heart: This is "the perfect use of the technology...Patients get the same kind of care they'd get if they were sitting next to me."[8] Colleagues at the Watauga Medical Center in Boone, North Carolina, use videoconferencing to analyze difficult cancer cases, conferring—via a network of computers, video, x-ray equipment, and a 60-inch color monitor—with oncologists at the Carolinas Medical Center in Charlotte 150 miles away.[9]

it has shaped new relationships between people and their doctors. Both patients and caregivers can use the expertise of specialists without the expense of travel and office visits. Diagnoses or informed second opinions become much more practical options: less expensive and more manageable. With the advent of the Infocosm, the constraints of location—on health care givers, their customers, and the kinds of facilities in which they work—will disappear.

CARE AT HOME

Confined to a wheel chair, Dorothy Ditmore sits at home in rural Minnesota and receives her health care through a computer hookup, which helps to keep her at home rather than at the hospital. She uses a two-way interactive system to be in touch with her care team who can do assessments, walk through standard tests, or monitor her heart.[10]

Consider such innovations as the "corner store." This concept moves medical testing equipment into local communities and links people to specialists at remotely located care delivery facilities. Altering our perceptions of place—*where* health care is delivered—will determine the type of caregivers who will be part of a person's care team. For example, one innovation is a home-care robot that is monitored by health care professionals in a central location. "HANC," as it is called, uses twenty-four-hour data, voice, and video transmissions to monitor vital signs, medication scheduling, and self-care training.[11] Innovations like that vastly reduce costs and improve quality of life for the sick and elderly—indeed for all of us. Anything that can be digitized—voice, medical notes, ultrasound readings, x-rays, pathology slides—will be.

And the digital information will transcend the barriers of both place and time to be used whenever and wherever people need it. And remember, digitized information travels across oceans and national borders as easily and quickly as it travels across town.

Form and the Infocosm

The transformation of atoms into bits—"being digital" in Nicholas Negroponte's parlance—is the essence of technology-enabled freedom from the historic barrier of form. Of the three traditional barriers, that may be the most difficult to visualize. It's a concept that challenges many assumptions. But it's imperative to understand the dissolution of form and what it means, because its impact will be the most far-reaching.

Let's look first at an everyday example outside of health care. We are all familiar with the standard organization of grocery stores: dairy products here, produce there, frozen foods in one aisle, laundry detergents in another. To make shopping from a home computer possible, we can digitize information about merchandise and store layout and store it in a database. We can click into this virtual grocery store and see a fairly accurate representation of our local supermarket—cereal, sour cream, shopping carts, and all. We see here how information technology has dissolved barriers of place; in a sense, the store comes to us. But time is also dissolved because the hours of business have no limits, no one waits for a cashier, and we waste no time traveling to and from the store.

With a click of your mouse, you can reconfigure the arrangement of the store and reorganize the aisles to fit your preferences—according to dietary restrictions, ethnic food types, recipe ingredients, you name it. The possibilities are endless. The shopper changes the store's form. And with the proper communication links among retailers, there are even broader possibilities. You could bring clothing, appliances, books, and sporting goods into your personally redesigned "store."

Picture today's typical doctor's office: a receptionist, a computer terminal for appointments, and long rows of file cabinets containing medical records and other materials. This is "form" with a vengeance: papers everywhere, sticky notes coming unstuck and getting trampled underfoot. In short, the physical structure of the traditional medical record is a major obstacle to smooth, efficient health care. Introduce the proper use of information technology, and a new form of record emerges—one of bits and bytes, not paper and ink. And this digitized, or virtualized, record is manageable, transferable, easy to store, organize, and manipulate.

Yet information technology can do so much more than eliminate paper. In addition to virtualizing medical records, we can also virtualize all or part of key processes, markets, events, organizations, collaborations, customer interactions, training and education, products and services—even people! Whenever we make the transition from the physical to the virtual, we create new opportunities to eliminate costs and expand options available to consumers and providers. The Guardian Angel—a prototypical patient-monitoring system—is a forerunner of interactive systems that will allow twenty-four-hour observation of patients in their homes. Even the form of many kinds of surgery will change as lasers, scalpels, laproscopic instruments, etc. are guided directly by computers under the supervision of surgeons.

THE GUARDIAN ANGEL

At Children's Hospital in Boston, scientists are developing a device called a "Guardian Angel." It will continuously monitor a patient's vital signs and alert the caregiver to changes in physical condition. If, for example, an incorrect dosage of medicine is prescribed, the machine will signal the danger and thus prevent the mistake from becoming lethal.[12]

Those higher-order forms of virtualization are image-based and multimedia in nature. Electronic Data Interchange (EDI) evolves into electronic knowledge collaboration and electronic commerce. In health care, electronic medical records, telemedicine, electronic databases of medical knowledge, virtual models of patients, and electronic training of surgeons will form the building blocks for new, virtual capabilities involving organizations, major processes, and markets.

Telemedicine offers numerous examples of how technology frees us from the constraints of health care delivery's traditional forms. While no machine can replace the benefits of human contact, we will soon have to redefine what we mean by the very word "contact." Support groups have sprung up all across the Internet; newly diagnosed patients and their families use the Internet to converse with others in similar situations. They share their fears, hopes, and experiences.

This is neither contact nor care in the traditional sense, but doctors are impressed with the results.

Already we see physicians and nurses delivering new forms of care, even when their customer is not physically present. An early example of such care is today's toll-free 800 service offered by health care givers who answer questions posed by callers who prefer not to identify themselves. Similarly, a fledgling teleconferencing system has started up in Boone, North Carolina. Boone's Watauga Medical Center hooks up with the oncology team at the Carolinas Medical Center in Charlotte and, in doing so, demolishes the traditional form of conferring with colleagues.

SUPPORT FROM THE INTERNET

"Cyberspace is brimming with millions of pages of medical information and thousands of 'bulletin boards' dealing with specific illnesses, from the common cold to Parkinson's disease." And here, Karen Caviglia and millions of others sit at their keyboards and get support and get educated—by people from around the world.[13]

In addition, the "form" of surgical procedures will change as multiple technologies allow both observation and participation from distant sites. In the past, to have a class in procedures, a medical student had to be enrolled in a course in surgery and visit the operating room. With the advent of the Infocosm, that will no longer be the case. Now the student can observe—and even participate in—the operation: Virtual reality allows a student to repeat an operation as many times as it takes to get it right.

Patients will also learn more about their own medical conditions and will use information technology to monitor their health following a distinctly new approach. Using virtual reality technology, caregivers can treat people for phobias in threat-free environments. For instance, by creating "virtual" environments, people can overcome a fear of heights without actually ascending an Eiffel Tower or a Space Needle. Such treatments suggest exciting possibilities for people who aim to lead more productive lives without spending years in expensive therapy.

New Assets Created by Information Technology

No one in the health care industry today will be fooled by claims that promise to solve all of our problems and take care of the ever-growing complications and demands placed on the industry. In the health care delivery system, as in most industries, there are many variables and unknowns. But there are specific, measurable gains which will emerge as information technology dissolves the barriers of time, place, and form.

Traditionally, the business assets that mattered were properties, supplies, franchises, and capital. The health care enterprises of the future, however, will measure their assets in much broader terms. We group these assets into four categories: speed, agility, reach, and insight.

Speed. Information technology obviously speeds up the delivery of care and the flow of information. But speed becomes an asset when a business has the intent, information technology, people, skills, and practices consistently to capitalize on the value inherent in quickly flowing information and fast-changing conditions. It gives health care professionals quick access to, for example, outcomes and treatment information, and thus enables real-time decisions in times of medical crises. In the heart of coal country, West Virginia, a doctor plans to post medical records on the Internet so that when needed, emergency workers will have instant access to such critical information as drug allergies and other chronic conditions.

Because technology can synthesize information, health care givers can deal simultaneously with multiple demands. Service will be faster, more effective, and more efficient.

MEDICAL RECORDS ON THE INTERNET

In Wayne, West Virginia, an economically depressed town of about 2,000, Dr. Bruce Merkin plans to record notes from patient visits electronically and put them on the Internet. Then, when one of his patients arrives at an emergency room on a Saturday night, doctors there will be able to gather essential information rapidly about the person's medical background. While it is a small-scale experiment, Dr. Merkin says he is convinced that "the Internet is the way to go."[14]

The benefits are clear. Much of effective health care delivery depends upon timing. Think of the emergency room, the delivery room, organ-transplant surgery, and countless other procedures. With quickly available medical records and instant access to specialists and other doctors, we can assure individuals much more fluid care.

Advances in information technology create the asset of speed in addition to speedy delivery of care. Any use of information in the "back office" processes of the business, for example, processing memberships in managed-care organizations, fielding queries and addressing issues, or processing claims, benefits from speed. Speed of processing also means that greater numbers can be served. By serving individual customers faster, organizations can respond to more customers in the same amount of time.

Agility. Many consumers are confused about health care today. As politicians, insurance companies, and care providers themselves bombard the general public with strident and often conflicting messages, people's attitudes about health care fluctuate wildly. Opinions run the gamut from anger and fear to worry and frustration. Nonetheless, the general public has developed a greater awareness of its own health care needs and the sources for health care delivery than any other generation. And greater public awareness means more demands for service from consumers, politicians, and employers. To be competitive and relevant, successful health care organizations must be able to adapt to new demands, fit the needs of shifting markets, and answer the demands of communities with unique health concerns.

Success in the future will increasingly be based on an organization's ability to continuously adapt to an ever-changing environment. Organizations and entrepreneurs who can continuously execute new strategies in the face of new challenges and opportunities will gain a new competitive advantage. Therefore, organizational agility is emerging as a new basis of competition. Information technology enables organizations to process and store larger amounts of data; smart software (expert systems, neural networks, genetic algorithms, etc.) helps us create insight and knowledge. When organizations train people and design processes to anticipate change and respond accordingly, we'll be playing in a new arena of organizational agility.

Reach. Consider a health care administrator searching the worldwide markets in Manila, Frankfurt, and Tokyo for the best prices on medical supplies. Imagine specialized care delivery organizations—for example, ones that treat diabetics—reaching across many national boundaries to provide advice and care to people around the world. This is the information technology-enabled asset of reach—the ability to access know-how, resources, and markets, wherever they are located. In tomorrow's health care industry, the reach will be global, and the connections must be established between all the resource suppliers and those who use the products. Our reach, in other words, will extend to those who supply not only medical equipment but also the capital, the expertise, and the information and knowledge—wherever these are created. Thus, reach

AUKERSHUS COUNTY, NORWAY

In Norway's vast Aukershus County, a seamless communications network links public hospitals, primary care physicians, health centers, social service offices, and pharmacies. Each year, the network is used to provide more than half a million people with comprehensive, integrated care and saves providers thirty to forty staff-years of paper work, travel, and administrative detail.

extends to expertise, knowledge, resources, and consumers who are also reaching toward health care resources and information for themselves.

37,000 SQUARE MILES OF HEALTH CARE

Combining personal computers, videoconferencing, and high-speed data transmission, the Eastern Montana Telemedicine Network gives rural residents access to physicians and rural physicians access to specialists. "One hundred percent [of the patients] said they would prefer telemedicine to travel."[15]

Reach is a critical asset to people who are geographically distant from major medical facilities; consider the plight of a Norwegian farmer requiring a consult with a specialist in Oslo. Reach also helps alleviate the isolation inherent in a rural doctor's job. Vast networks have already been created in Aukershus County in Norway, eastern Montana, Hawaii, and other areas. Although the waters of the Pacific may have spurred Hawaii to capitalize on the information technology-enabled asset of reach, others are working to achieve the same results without the geographic barrier.

Obtaining unique capabilities from multiple sources is one innovation that is made possible by the extended reach provided by information technology. When health care organizations engage in competition of this sort, the opportunities to provide better service at more efficient cost change the rules by which the industry operates.

Insight. The ultimate asset in the Health Care Infocosm is insight—the ability to access, filter, analyze, and synthesize diverse information and experience to allow us to see new patterns, new relationships, new ideas. Of course, insight is not new; but we are talking about purposefully pushing to new levels of insight as a mechanism for organizational growth and vitality. As George Halvorson suggests in *Strong Medicine,* knowing what questions to ask is an act of creativity as important as finding the answers.[17] This paradigmatic shift is the most significant of all the changes that we describe here.

HAWAII GETS WIRED

"Hawaii has been rather isolated," says Janice Kato of High Technology Development Corporation, but with new information technology "the world is getting smaller." Fiber optic lines and greater bandwidth allow telemedicine connections between Oahu and Molokai, so that caregivers can now share a wide range of medical information electronically. "Today we're using [telemedicine] to treat patients. Forty years ago, we were using telephones," Kato says.[16]

Citing a study done by Gray, Capone, and Most in the *New England Journal of Medicine* in 1991, Halvorson points to a case of a Rhode Island hospital where 185 patients became comatose after suffering heart attacks. Each was admitted for emergency treatment and later moved to intensive care; each one was given an identification number and assigned to a group of health care professionals who kept the appropriate records for treatments and outcomes. The hospital knew the outcome for each individual, but no one ever made it a point to ask what happened to these people as a group—no one, that is, until a team of independent researchers

MEDICAL JARGON ON INTERNET

Thirty-thousand people a day visit a site on the World Wide Web portion of the Internet to demystify medical jargon. Allowing health care consumers to translate "myocardial infarction" to "heart attack" provides them with the answers they increasingly demand about their medical conditions and possible treatments.[19, 20]

sought to determine the survival rate for those patients. Halvorson reports that, unfortunately, they all died. But he is most astonished by the fact that no one knew this, simply because no one had insight into the overall situation. Nobody asked the right questions about the group. "These 185 consecutive deaths were invisible as far as the health care system was concerned. They were individually reported and then simply forgotten."[18]

Information technology enables us to know what questions to ask by shaking us out of our traditional ways of thinking about health care delivery. Without the right questions we can never reach meaningful insights. It even helps us make sense of the question by translating medical jargon into laymen's terms. Health care organizations already recognize the added value that people with insight bring to the delivery of care. The organization that can take raw data and turn it into sharable information positions its people to create new medical insights and knowledge.

Forward into Tomorrow

The vast and rapid changes that information technology brings to the health care industry make it clear that we can no longer afford to do business the old-fashioned way. Assuming "business as usual" is likely to be fatal in a world where time stands still, everybody is in the next office, and work can be transported 8,000 miles in a quarter of a second. Sound crazy? These are the potential implications of information technology-enabled freedoms from time, place, and form. Innovations in information technology can change the way that health care professionals think of themselves and their work, altering the relationship between the health care facility or organization and the individual (who is just as likely these days to be referred to as the "customer"), forcing the health care enterprise to assess its core competencies and focus its efforts on what it does best.

Health care organizations cannot afford to remain isolated and to think that these waves of change will wash by them, leaving them unaffected.

http://nytsyn.com/medic

News sources from Asia, Europe, and the United States contribute features to an Internet service called "Your Health Daily," a site on the World Wide Web that is offered by the New York Times Syndicate. On-line users can now receive real-time, authoritative health and medical news. The site covers everything from the latest heart disease research and new reports on asthma, to AIDS conferences and features on nutrition on health.

Making full use of information technology will be expensive, both in terms of dollars and jobs, but the organization that refuses to participate or that goes about the process halfheartedly will likely find itself out of business.

Unfortunately, most health care organizations have been slow to adopt innovations. While commercial information technology initially focused on structured data and text, the health care industry required integration of other forms of information—for example, video, audio, images, waves, and scans. This may explain, in part, why health care has lagged behind other industries in exploiting information technology. But the roots may lie in organizations themselves. Many are concerned about rising costs or fearful of change. Neither can be avoided, although navigating a course through them can be made much easier if the organization is willing to assess its functions and structure honestly.

We believe that this assessment will encourage the business to undertake a transformation of the enterprise, a radical and rapid response to changes in the health care industry. In the next chapter, we describe how such a transformation enables an enterprise to move from where it is to where it needs to be and, accordingly, how it can survive in the health care environment of the twenty-first century.

CHAPTER 2

GETTING FROM HERE TO THERE

StV = (3V 5A 10/10W)

This is both a formula for success and a road map for a journey. Dr. David Campbell, CEO for St.Vincent's Hospital in Melbourne, translates this equation as follows: St.Vincent's will be the best hospital in the state of Victoria in three years, the best in Australia in five years, and among the ten best in the world in ten years as measured by standard industry benchmarks.[1]

That was a startling, if not boastful, prediction, given what Dr. Campbell saw when he first came to St.Vincent's in 1991. The hundred-year-old hospital, founded and directed by the Sisters of Charity, was crippled by inefficient care delivery processes, a highly fragmented organizational structure, antiquated facilities, and outdated information systems. The result was poor patient service, high operating costs, and a doubtful future. According to an internal study, caregivers spent just a little over *half* their time actually caring for patients, while hours were wasted in administrative and logistical tasks.

To turn St.Vincent's into a modern, efficient, and patient-focused health care organization, executives enlisted the support of the entire staff and began a transformation literally from the ground up. The St. Vincent's personnel reengineered care processes, established multidisciplinary care teams, consolidated job descriptions, and developed state-of-the-art information technology capabilities and facilities to support the new operating model. In the process, the staff reduced costs—by 18 percent—and improved patient service significantly. For example, they shrank admission times from 45 minutes to 5 minutes, and all of that was accomplished in just over two years (1993–1995).

Few health care industry executives, even five years ago, imagined the potential of transformational change to dramatically strengthen the delivery and management of health care. Only the rare visionaries, like Dr. Campbell and his colleagues, perceived the dim outlines of the promise ahead: transformation of nearly every aspect of the health industry. Likewise, only a few visionaries would have spotted many other factors: the tremendous benefits possible through information technology, community-networked organizations, noninvasive treatments, and opportunities for worldwide collaboration with specialists. Innovations in information technology are rewriting the rules by which all players plan their strategies, organize their business processes, and marshal their forces.

Those organizations whose executives have taken note of the new technologies' industry-shaping forces are the organizations that will secure positions in this rapidly evolving game. In this chapter, we examine how to harness those forces and their impact on health care organizations that are trying to come to grips with simultaneous changes in strategies, processes, information technology, and people. As we saw in Chapter 1, remarkable advances in information technology promise incredible benefits but at the same time demand that organizations embrace technology as an integral part of their business formula. The organizations that are successful in the new environment of the Health Care Infocosm will be those that can adapt rapidly and constructively to the changes.

Changing All and Changing All at Once

The challenge of the Infocosm is comprehensive: In the age of the Infocosm, health care organizations must reassess and redesign—strategies, processes, technologies, and people—completely, and they must bring about the transformation of every part of their organizations simultaneously. The integration of changes among these components is illustrated in Figure 2-1. The health care organization in the Infocosm has to make

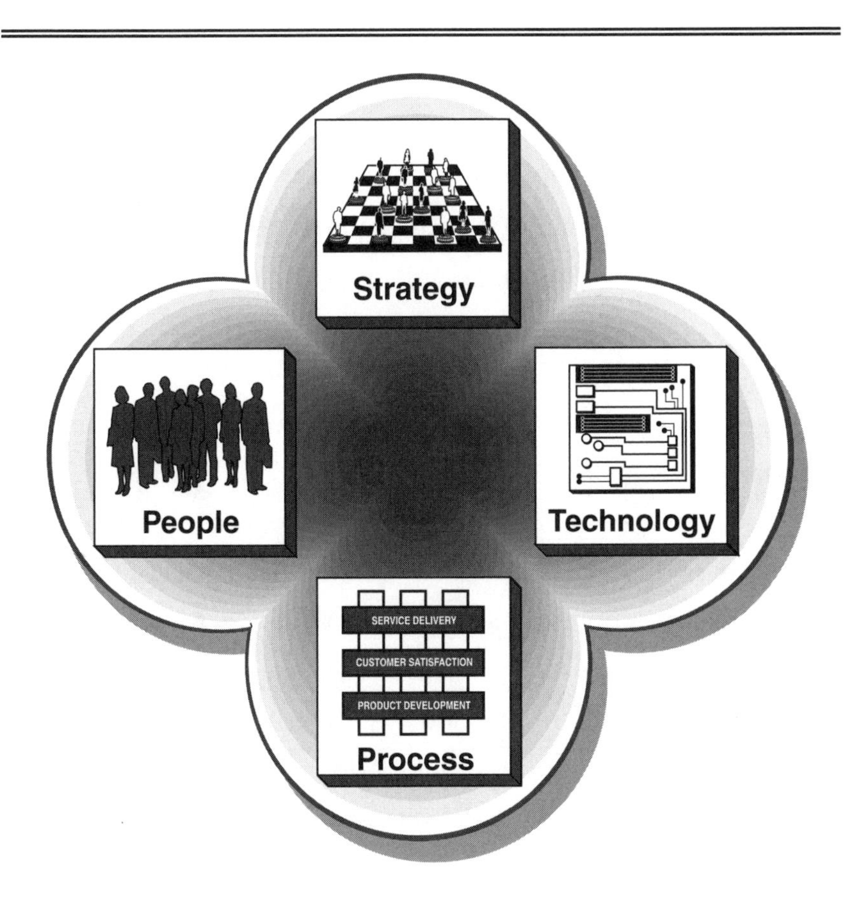

Figure 2-1. The Components of Transformational Change.

changes to *all* of these parts of its enterprise. Furthermore, it must do all of these things *at the same time.* Another way of putting this is expressed in the well-worn metaphor for imagining the impossible, which forms the question many executives ask: "How can I build this ship while sailing it?" Given how much flux the industry is now experiencing, we could refine this question by asking, "How can I build this ship while sailing it on turbulent seas?"

New entrants from other industries have joined the health care arena, which grows increasingly international in its coverage. The Health Futures Forum in Australia, for example, invites experts from the medical professions, public and private health care organizations, government agencies, and the telecommunications industry to join together to chart the possible future of the health care industry. Those experts envision advances in information technology that will fuel worldwide competition among businesses in this increasingly integrated industry.[2]

But embracing the promise of the Infocosm is hardly more than a start. Every population of every organization is understandably stuck in patterns built from years of assumptions and activities. Rare indeed is the person who doesn't regard the unknown with some trepidation. Health care executives themselves face a daunting future: Convince staff to strive for progress, to make the most of new organizational structures, to use information technologies in innovative ways that create new knowledge, and to encourage collaborative attitudes toward the wholly new ways of managing and delivering health care. Fast-changing information technology, in short, is destabilizing the industry while it enables all players to respond creatively to its evolving rules.

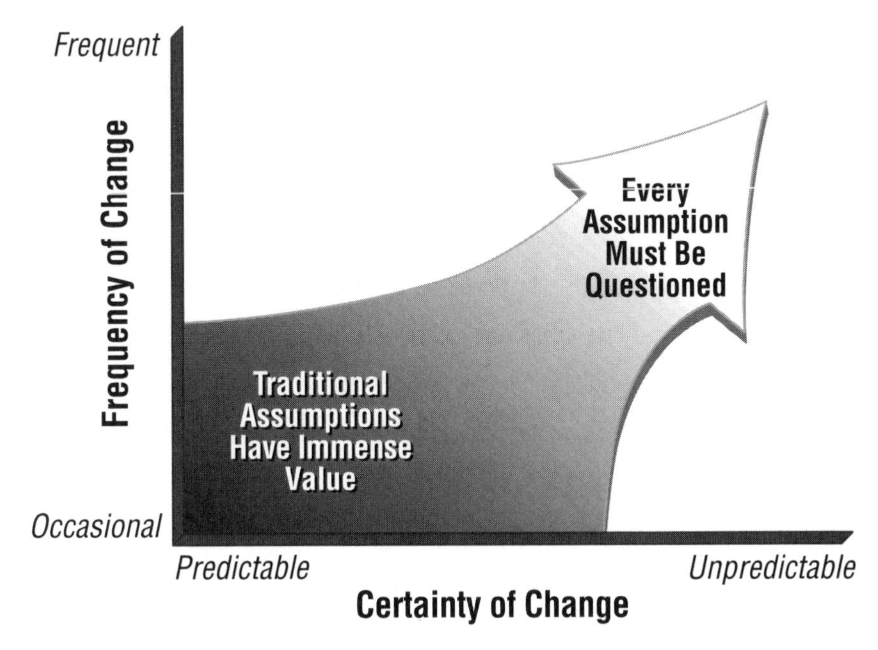

Figure 2-2. Challenging the Old Assumptions.

Transforming an enterprise means challenging the assumptions that made it a success in the first place. In stable times, change is relatively predictable, and the organization can focus on achieving efficiency and productivity goals. In times like those we experience today in the health care industry, change is both frequent and unpredictable; thus, all of our assumptions may no longer be—and probably are not—valid any more. This is illustrated in Figure 2-2. When customers discover that they have more choices, health care executives find that they must continually question the old assumptions, even one as fundamental as the boundaries of their existing organization.

This turbulence invites us to imagine a future that many might consider impossible: health care enterprises transformed to set new strategies, design new processes, and define an entirely new relationship between people and their organizations. Nevertheless, our aggressive outlook for future core business processes that extend across organizational boundaries, information technology architectures, and the structure of people's jobs and their organizations implies a dramatic and rapid transformation of the entire health care enterprise. Accordingly, information technology, coupled with all the other changes that comprise an enterprise transformation, will create new imperatives for the industry: to make deep shifts in values, work styles, attitudes, skills, and behaviors.

Transforming an enterprise, however, is not synonymous with "reengineering," as Michael Hammer and James Champy defined that term[3] or—as it has recently come to be understood—as primarily a redesign of business processes. The transformation we anticipate integrates strategic planning, information technology, process design, and changing behavior throughout organizations.[4] The transformational journey must also be carefully planned and orchestrated, representing "execution as strategy."

Andersen Consulting hypothesized that an enterprise transformation of such magnitude and scope would demand a broad vision and comprehensive responses to change imperatives. To support that theory, Andersen Consulting investigated and mapped the journeys taken by more than 50 multinational corporations that have undergone such major changes. Those companies include British Airways, Direct Line Insurance, AT&T, Chemical Bank, Levi Strauss & Co., SmithKline Beecham, and Xerox. Another source of understanding such deep transformational change came from Astra Merck Inc., a brand new, billion-dollar pharmaceutical company spun off from a joint venture of Astra AB of Sweden and Merck & Co., Inc. Astra Merck provides both a model for building an ideal organization and invaluable lessons about orchestrating simultaneous change in strategy, people, technologies, and processes.

Each of those companies demonstrated that to transform itself successfully, it had to develop and articulate its own strategy for change. Furthermore, its strategy had to fit with the scope, pace, and direction of current realities, both within the organization itself and in the marketplace at large.

We believe that to postpone action is a perilous option. Winning organizations are those that accept the challenge, and those that tarry will find themselves playing a virtually impossible game of catch-up. As a longtime health care executive and CEO explained to us, "The very life of the organization depends upon the ability of its leaders to get the organization where it needs to be for an unforeseen future."

There are, of course, executives who, recognizing that their organizations already face a shaky future—shifting markets, falling revenues, aggressive competitors armed with the latest technologies, and changing demands of customers—have responded to those specters with strategies that cut wide swaths through organizational budgets while people in the organization continue doing the same things as before, only with fewer resources. The processes themselves have not changed: The pharmacy, which still dispenses prescriptions, has to maintain a steadily growing volume with fewer pharmacists. Intensive-care nurses remain responsible for the same paperwork, but there are fewer nurses on each shift.

Physicians still examine patients but, with less support, they have to cut short their time with each one.

Should improving economic conditions bring an increase in revenues, organizations that had cut their budgets in the name of transformation gradually resume their spending patterns, achieving neither permanent cost reductions nor organizational innovation. It is obvious that such patchwork reactions to difficult times seen in industry after industry can never yield grand results. That pattern is reminiscent of dieters who easily shed five or ten pounds in the first few weeks, only to regain the weight once motivation fizzles.

We believe that to achieve viability in the long term, executives must subject every aspect of their organizations to examination. It may be necessary to rebuild every process, every information system, every job description, and every relationship within the organization and beyond its walls. To illustrate just how radical a transformation can be, we present the case of El Dorado Health Care, an acute-care facility in Alta Vista, Arizona, which in August 1993 undertook such a metamorphosis. Although El Dorado is a fictional organization, we base its story on the experience of a real acute-care facility that transformed its strategies, processes, information technology, and relationship with its own people and with the public.

The Arizona Miracle

Picture the staff's reaction to its leader's announcement that the organization's annual operating budget had to be cut by 50 percent. What's more, that leader insisted, quality of service must remain high and, in certain cases, improve. How would people respond? Dismay, panic, incredulity, and snickers would surely characterize the atmosphere—at least for a while.

The CEO's announcement was actually more of a battle cry than a mere suggestion to cut expenditures. The CEO was Sandra Smith, and she had seen the forces amassed against El Dorado. When she presented her vision and strategy to change utterly every aspect of El Dorado Health Care—both the way El Dorado was doing business and the very business itself—her staff greeted her plan with understandable apprehension.

According to Smith, her decision to move El Dorado from its previous posture to a position of improved capabilities was not optional: It was imperative. The facility wouldn't survive in the economic environment—a commodity market driven by prices and by strong competitors—that had developed in the Alta Vista area. In earlier years, El Dorado Health Care had built its reputation for superior care in oncology and orthopedics. It had developed those as centers of excellence. Nearly forty years old, however, El Dorado was hampered by its antiquated, low-level technologies—both medical and information. It couldn't continue to provide quality services without incurring debt or otherwise financing its investments in new technologies.

In 1993, forecasters predicted that El Dorado's inpatient occupancy rates would fall 10 percent each year for the next three years, while emergency room visits would climb by 7 percent yearly. By 1997, outpatient services would account for 85 percent of the their total revenues. El Dorado, in the hot sun of Arizona, was on a burning platform.

This is not to say that El Dorado was without assets. It had strong ties with a nearby university and a range of neighboring health care facilities, and its staff of young, energetic physicians had earned it respect for its personalized care.

Smith and other executives at El Dorado recognized that they needed a transformation that was both fast and comprehensive. Her plan called for a radical change in strategy. El Dorado would establish new kinds of partnerships with other health care organizations, design processes that would replace age-old functions, redefine their services, and initiate an extensive overhaul of information technology.

Strategy

Strategic intent is like a marathon run in 400-meter sprints. No one knows what the terrain will look like at mile 26, so the role of top management is to focus the organization's attention on the ground to be covered in the next 400 meters.

—Hamel and Prahalad[5]

An enterprise transformation is a mission of monumental proportions, and the hallmarks of a successful transformation are a well-planned strategy and effective communication of that strategy to everyone in the organization. The executives who set the strategy must anticipate the direction markets are taking.

Predicting market trends proved a speculative challenge. For example, no one could forecast the direction or speed of the federal government's health care reforms that seemed imminent in August 1993 when El Dorado began its transformation. Moreover, the local economy presented its own set of variables: Alta Vista was in a mature managed-care market, with penetration in the 50 percent range. Consolidation of the players in the market had already begun, leading to increasing competition for the area's two million residents. The consolidation and increased competitiveness were already starting to strain the resources of the more than two dozen hospitals serving the area.[6] Market strategists estimated that by 1998, as much as 70 percent of the local population would be covered by managed-care contracts and that inpatient occupancy rates would plummet in all area hospitals. All members of the El Dorado team agreed that the effects of those market forces presented a high degree of inevitability.

Furthermore, Smith and the El Dorado management team realized that to bring about the necessary changes, El Dorado needed to upgrade its current information systems. But no one could be certain that the hospital would be able to develop the information technology components fast enough to meet the deadlines of the transformation. The executives also expected that during the transition period the hospital would lose many of its best employees. The most questionable aspect of the entire endeavor was, in fact, the staff: Would the physicians, the nurses, and the rest of the personnel support changes in every area of care delivery and support services?

Smith recognized that virtually every department, every organizational structure, and every employee at El Dorado would have to endure the anxieties of upheaval. She had calculated the impact of her announcement in both financial and psychological terms: To survive, El Dorado would have to cut costs nearly in half. She further concluded that the health care facility was extremely vulnerable to market fluctuations and that—in spite of the risks of undertaking a transformation—she had no alternative: "We might as well fail trying," she said, "than fail doing nothing. If we are going to go down, let's go down fighting."

She mapped an aggressive six-month master schedule for identifying all of the change that would be encompassed in the transformation. The team started to define their strategy in mid-August 1993. After they began, their mission shifted from a desire to position El Dorado as a stand-alone organization to a recognition that they would need to become part of an integrated delivery system.

The strategy-setting phase lasted approximately two months, from mid-August to early October 1993. Almost simultaneously, managers sought ways to redefine how *everything* was done. Interviews with health care personnel revealed that there were practically as many visions and strategies floating around as there were people on the payroll, and the transformation team realized that it would have to communicate clearly and frequently with everyone at El Dorado in order to make this a truly collaborative effort.

During that early phase, the team determined a set of principles to shape the mission statement and make the transformation happen. Heading their list was an uncompromising commitment to maintaining the best medical care, even in the face of the turbulence they expected during the transformation. Market trends indicated that health care organizations were quickly directing their focus on members' wants and needs. Although the prevailing image of members as consumers played a small part in this shift, the major impetus arose from the recognition that developments in information technology would dramatically alter the relationship between caregiver and patient. El Dorado aimed to exploit that evolving relationship.

Processes

Integrated systems will add real value when they can describe, not their new organization chart, but new integrated experiences for those they serve.

—Donald M. Berwick, CEO,
Institute for Health Care Improvement[7]

Among their most startling initiatives, El Dorado executives took all 84 of the hospital's core and supporting departments and revamped them into seven specialty care centers: critical care, acute care, emergency, transition, oncology, outpatient treatment, and wellness and prevention. Seven support centers—financial, clinical, human development, facilities, technology, stewardship, and business development—were established to buttress each care center. That realignment alone eliminated just under half of El Dorado's departmental work. By focusing on patient-oriented processes, El Dorado removed bureaucratic waste, streamlined its business functions, and consolidated the work of various departments into care centers responsible for managing care delivery.

El Dorado set up three of the seven care centers in 1994.

One, the Acute Care Center, took patients who had previously been assigned to other clinical areas. Structured around teams of health care professionals who focus solely on their patients' well-being, the Acute Care Center defined partnering as their new standard for bedside care. Three-partner teams take responsibility for delivering the care: a nurse partner directs the care, administers medication, assists in discharging patients, and performs other duties as the need arises; a multiple-skilled partner, or MSP, monitors vital signs, performs EKGs, administers basic patient care, and keeps charts up to date; and a patient-service partner takes responsibility for transporting patients, serving meals, cleaning rooms, and helping nurses. Patrick Russell, director of nursing at El Dorado, has high praise for partnering. He notes that shifting responsibility to the Acute Care Center has helped the staff focus on the benefit of its patients rather than on an "abstraction called the 'hospital.'"

The Critical Care Center also uses the partnering approach to consolidate the work of various departments, which functioned independently under the old structure. With minimal disruption, hospital administrators relocated the radiology department closer to the critical care wing and constructed an information center where patients' families can meet with physicians and nurses, view videotapes about home treatments for recovering patients, and speak with hospital personnel about billing and insurance payments.

The new Oncology Care Center consolidates five former departments—each headed by a mid-level manager and in some cases assisted by a management specialist—into a single team of professionals with a single care-center leader.

The Oncology Care Center serves patients in every stage of cancer treatment—from initial diagnosis, through inpatient treatment, to outpatient visits for routine follow-ups. The new center's teams are self-directed and extensively cross-trained to allow the center to operate with fewer managers and health care professionals. The self-managed work teams have reduced the need for management full-time equivalents (FTEs) by nearly 70 percent. Because of their immediate responsibility for patients, the teams make decisions closer to the individual. Patients also appreciate that the center bundles all charges: Rather than sending itemized bills from the pharmacy, radiology, laboratory, and a host of other services, El Dorado now sends the patient a single per diem bill. The oncology support team can complete patients' charts in half the time it used to take, and now that all caregivers use a standardized form for documentation, those charts report valuable information more accurately and concisely.

By virtue of its FTE reductions and its team approach, the Oncology Care Center has reduced costs by 32 percent, provides better service to customers, and makes better use of professional staff.

When El Dorado managers turned their attention to the redesign of its business-support processes, it made sure to focus on the patient rather than on the hospital or its physicians. They sought ways to improve communication with patients, to consolidate treatments, and to minimize the time spent on activities not directly related to patient care. Their efforts have eliminated many clerical activities, which had consumed so much attention under the old regime. A major component of that transformation—the dramatically improved information management system—includes a solution that integrates multidisciplinary care paths, on-line charting, patients' electronic medical records, and order management.

Information Technology

Re-alignment of the priorities and structures of the delivery system is only possible if supported by a corresponding re-alignment of information technologies and information management processes.

—Young and Martin[8]

El Dorado's transformation called for a logical source of all types of information for caregivers and hospital administrators. While they realized that they might have to start with a partial solution, they ultimately wanted a fully integrated system to accommodate an electronic medical record for each patient, an account of multidisciplinary care paths, as well as charting and order management.

Smith realized that a paperless operation, with information accessible at workstations and on handheld computers, would be extremely valuable to caregivers throughout the organization. Executives aimed for a nonproprietary information system that would permit easy communication within the hospital itself as well as with suppliers, customers, and other health care organizations beyond its boundaries. The new information systems, for example, would need to allow caregivers to establish connections throughout El Dorado so that staff could participate in such activities as videoconferences and consultation with medical colleagues throughout the Southwest and elsewhere.

Recognizing that bringing El Dorado's modest information management systems would prove almost as disruptive as relocating departments or introducing collaborative work structures, hospital executives took pains to include staff members in every step of the transformation. The El Dorado team knew that caregivers at every level would benefit from hands-on demonstrations of the new systems.

El Dorado would need to train staff to use software programs that would give them instant access to various kinds of health care data: patients' medical records; studies of specialized procedures, treatments, and drug therapies; research in oncology and orthopedic specialties from major world health centers; and online connections to caregivers in health facilities in major metropolitan areas in the United States and abroad.

The information systems had to permit staff to capture data when and where it was generated and provide real-time notification of results when and where people needed the information: the laboratory, the emergency care center, the bedside. The systems should, for example, accommodate physicians' notes on a patient's progress and provide a chronological log of treatments and results in a file readily accessible to authorized caregivers.

But El Dorado approached that goal from a modest starting point. While the accounting department, for example, kept copious records for billing purposes, account processing often waited for weeks for clinical information to filter through departmental functions. The independent departments—much to the irritation of the patients—collected redundant information, and caregivers were frustrated with the incomplete collection of fragments that represented each patient's case.

Early in 1995, the hospital initiated the projects that will eventually produce the information systems the team envisioned. After reviewing several patient-care technologies, hospital executives knew they wanted both the flexibility and the reach to provide patients, nurses, and physicians twenty-four-hour access to medical information from anywhere. El Dorado's systems have already established connections with most of their providers, and the team has been able to implement its initial restructuring of processes. Even though the full benefits of an enterprise transformation may not be evident for some time, El Dorado—like many health care organizations—can start realizing gains early in the transformation process.

In fact, the internal communications systems that were put in place to achieve those connections—including E-mail, a caregiver call system, drug ATMs, an electronic patient status board, group scheduling, a CD-ROM library, electronic data interchange, and a much-improved telephone system—already allow caregivers at El Dorado immediate access to extensive information about patients, drug therapies, and the latest research in oncology. Each care center's network allows instantaneous data exchange among physicians and nurses, administrative associates, managers, and such ancillary personnel as lab technicians, radiologists, pharmacists, and dietitians. Physicians carry wireless tablets or use bedside terminals to record patient data, order treatments, prescribe drugs, or educate patients on surgical procedures or treatments. Personnel use the new system to communicate with banks and insurance companies about payments, with suppliers about inventory needs, and with employers about patients' progress.

Those communications capabilities position El Dorado for fast response to changes in the health care environment. The technology doesn't come cheap. However, the technology, especially the automated care paths, are the linchpins of the transformation. El Dorado's executives had both the foresight and the courage to commit to their course early in the process and they pursued each initiative relentlessly until they achieved success.

Donald Stevens, El Dorado's COO, credits management with the successes of the transformation. "Were it not for the wholehearted commitment and courageous leadership that the executive team showed," he says, "El Dorado today would be a relic of the past. The enterprise can succeed in its transformation only if everyone in the organization becomes involved in the effort and executives commit the necessary resources to information technologies now rather than waiting until absolute certainty can be guaranteed." Not only does information technology help El Dorado provide excellent care, but it also redefines the hospital's work and the way individuals do their jobs.

People and the Organization

The rules have changed (and continue to change). What used to be fairly clear-cut career paths are up for grabs, and anyone unable or unwilling to adapt is at risk of being left behind.

—Warren R. Ross, Editor,
Medical Marketing & Media[9]

Many people in today's business world have interpreted the reengineering craze as a thinly veiled euphemism for layoffs. While it's not a foregone conclusion that many will lose their jobs in transformations like El Dorado's, it is accurate to say that just about everyone will see his or her job redefined, realigned, or redesigned. El Dorado executives estimated that the process would affect the jobs of 98 percent of its people. Clearly, there was cause for alarm. On the other hand, as CEO Sandra Smith points out, if El Dorado had done nothing, *every* job would go.

When she announced the strategy to reduce operating costs by half, Smith intended to create a mechanism to initiate the broadscale changes and to inspire people at El Dorado to join in the transformation efforts. During the transformation, El Dorado reduced its workforce by 33 percent with concomitant reductions in expenditures. While those reductions were wrenching, El Dorado did transform its position in the community, providing better care, both efficiently and effectively.

Among the benefits of El Dorado's radical transformation is a shift in emphasis: from individual hospital workers with prescribed sets of responsibilities to self-directed teams of workers. Working on one's own, a missing skill might go undetected. The new structure, however, encourages physicians and other caregivers such as nurses, therapists, and technicians, to consult, collaborate, and complement one another's capabilities.[10] Each employee redefines primary duties and distinguishes them from shared duties.

El Dorado's team reduced the number of organizational levels from seven to four. At the top of the new hierarchy, Sandra Smith, the CEO, continues to oversee hospital administration as she did under the old regime. All other managerial positions have changed. The Center Leaders from the fourteen new care and support centers replaced—as we noted above—eighty-four department heads and other management supervisors. At the base of the organizational structure, the staff members of the centers replaced employees assigned to various departmental functions. Decision-making in the transformed El Dorado now occurs at the edges of the organization—close to the community and customers it serves. According to the director of ancillary services, the El Dorado transformation differs markedly from changes at other hospitals around the country. When El Dorado created its seven care centers, the team regarded no area of the hospital as a sacred cow: Every job—even the CEO's—was reviewed, and everyone in the hospital felt the profound effects of the transformation.

Certain staff have found the transformation of El Dorado understandably frustrating. As radical as the changes are, and as rapidly as they are being implemented, many employees would prefer a faster and less extended process. At one point, when part of the hospital was still operating according to the old rules and another had already been transformed, people struggled to establish clear lines of communication. Linda Newman, a nurse in the oncology unit, said, "It's very painful that we have to go through this transformation, and yet I want it to be over. Let's get to the new world faster."

Although the team anticipated resistance, it occasionally arose from unexpected sources. The nursing staff, for instance, preferred the traditional designation *registered nurse* much more than the new title *nurse partner.* Although that seems a minor complaint, in the face of many disquieting organizational changes, such resistance indicates the depth of strongly held values and attachments. When, at last, every title in the new care centers ended in the word "partner," every employee felt the force of a great equalizer.

El Dorado executives learned valuable lessons about what it means to say that teams of people are "self-managed" and about how long it takes for people to take charge of themselves. Donald Stevens reports that the executive team incorrectly assumed that the hospital could achieve self-management in the first one or two years. That assumption proved to be unrealistic, and the team has since revised its estimate to five years. Stevens exclaimed, "We thought we were being conservative in our estimates!"

Not everyone relishes the idea of managing oneself, assuming ownership of the processes, and taking responsibility for decisions that under the old system were left to the "boss." Whether it's a desire to avoid conflict or an unwillingness to participate in the transformation, many at El Dorado—and organizations like it—prefer old patterns. Thus, any effort to transform an enterprise must incorporate a strong educational effort to help employees learn how to operate in a new world of team efforts, collaborative work styles, and self-management.

Herein lies one of the great paradoxes of transformation: Until an organization achieves self-management, it cannot easily cut managerial levels. While we expect people to learn to manage themselves, we must still keep the "boss" in place. During the transformation, people may leave because they lose patience with the new order or because their resistance to the system overwhelms their commitment to the vision. Sandra

Smith attempted to avoid such problems by involving every employee in many ways throughout the process. Her team developed a communication plan that included a hotline, articles in the El Dorado newsletter, and chances to speak with employee groups. Members of the team met with employees to explain the reasons for the transformation, to answer their questions, and to keep them informed about the hospital's progress. No employee was allowed to feel left out of the process.

One Journey, Many Paths

The road up and the road down is one and the same.

—Heraclitus

No one at El Dorado in mid-August 1993 could see exactly where the transformation would lead. Once it began the transformation, people at El Dorado discovered that they couldn't go back to the old way of doing things. No one could say, "Oh, I really didn't mean to take this path toward the goal!"

We don't present El Dorado's transformation as a model to replicate. Although El Dorado is exemplary, its results are not exactly reproducible. What is most significant about El Dorado, however, is that its leaders had the foresight and the courage to propose and to implement remarkable change—even before they knew precisely which paths would take them to their destination. Today's health care environment can't provide the luxury of stability. An enterprise transformation will inevitably introduce, at least temporarily, even more disruption as the organization seeks to change everything—strategy, processes, information technology, and its relationships with people—all at once.

Still, enterprise transformation is not a step off a cliff into a dark abyss. As El Dorado's example illustrates, corporate leaders start with an understanding of the landscape, and they evaluate their assumptions: As we survey the landscape, what facts can we observe and agree on? Do we all hold the same, or at least compatible, assumptions about what the facts mean for our organization's survival? What end do we desire, and can we agree, at least initially, on a course of action to help us realize that goal?

The Greek philosopher Heraclitus remarked somewhat cryptically that the way up and the way down are one and the same. Getting from here to there, getting our organizations to where they need to be, can take numerous turns and twists, few of which are predictable at the outset. As we demonstrate in the next three chapters, such journeys never trace straight lines from point A to point B. Rather, we chart our courses in multiple areas simultaneously: We need to develop a strategic vision and intent, redesign the processes that currently occupy the health care industry, put new technologies to work, and define the new ways people and organizations will work together. Only by engaging wholeheartedly in the transformation will we be able to realize the exciting possibilities of the new order of the Infocosm.

Chapter 3

THE JOURNEY BEGINS

Even when most welcome, change can be the source of seemingly endless disruption and frustration. Experiencing the effects of that dislocation, Linda Newman, an oncology nurse at El Dorado Health Care, expressed the feelings of many of her colleagues when she exclaimed, "Let's get to the new world faster."

Transforming the health care enterprise is, however, an extensive and complex process. No matter how disruptive change is, it need not immobilize people. Properly managed, the journey of enterprise transformation can follow a path in spite of the understandable uncertainties. The key requirement is focused, careful preparation that aims for a vision of shared goals and intended destinations. Until all relevant parties agree on both where they stand now and where they're headed, they will make no significant progress.

In this chapter, we'll discuss how to engage in strategy-setting that is consistent with the fast-moving environment of the Health Care Infocosm. The contrasts of strategy contexts is illustrated in Figure 3-1.

In the past, strategies used to focus on products and markets. Assessing those forces was an annual exercise, primarily the responsibility of executives and directors. Strategies for today and tomorrow, however, must be solidly grounded in organizational core competencies. The assessment is a continuous process in which everyone in the organization participates.

Strategy	YESTERDAY	TOMORROW
	Product / Market-Based	Core Competence Based
	Annual Exercise	Constant Vision
	Board Defined	Employee Owned
	Linear	Iterative
	Execution Focused	Context Focused

Figure 3-1. Comparison of Business Strategy Contexts.

Business plans that direct the execution of sequential activities will no longer suffice. New strategies must be fluid: They must change as the context changes.

Four essential elements health care managers must examine as they attempt to get their organizations on the road to increased productivity and competitiveness in the Health Care Infocosm form the framework for successful organizational action: reading the business landscape, evaluating shared assumptions, constructing an operational vision, and setting a course of action. Figure 3-2 illustrates those elements. From the broadest view of the business landscape, organizational action begins with an analysis of assumptions to a definition of the operational vision and, finally, to the optimal course of action to achieve enterprise transformation. Although our discussion will consider them sequentially, the elements work together in a continuous cycle.

These are the four elements in the visioning framework:

- Managers must take a realistic survey of the *business landscape.* Their goal is to chart the industry's current position and their position relative to that, taking into account those trends and conditions likely to cause upheavals in the future.

- The managers and executives must develop a set of *shared assumptions* about the implications of those trends and conditions. Naturally, the managers and executives involved in the transformation will come to the process with different beliefs and values, and they must either acknowledge and address those issues or risk sending conflicting signals to the staff they wish to turn their vision into reality.

- The managers and executives must craft an *operational vision.* Far different from traditional business planning, which introduces incremental shifts in resource allocation and reasserts a vague desire for prosperity, this organization-wide mission clarifies purpose, inspires change, and defines critical success factors, value to stockholders, and performance targets.

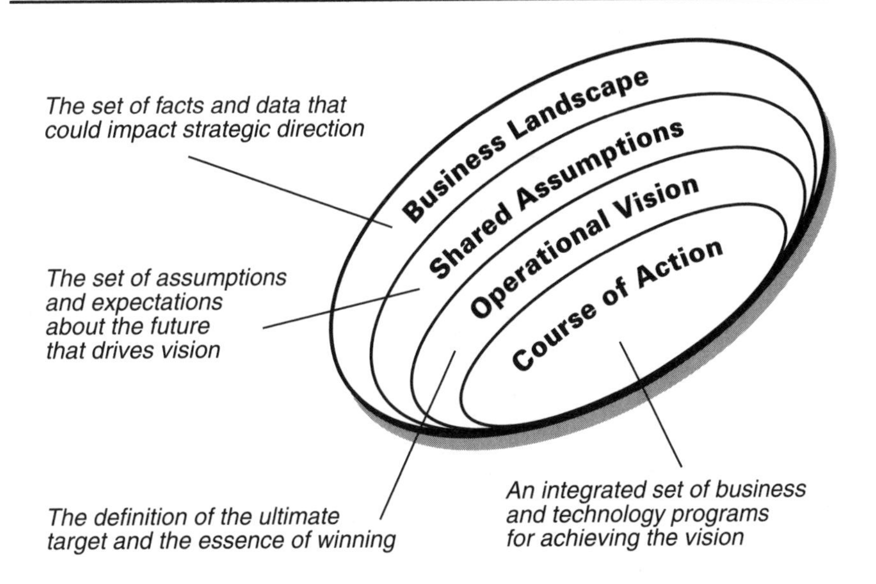

Figure 3-2. Visioning Framework.

- The managers and executives must determine a *course of action* that embraces information technology as a vital means to the desired business end and reflects the alternatives the group considered as well as the path the group chooses to follow.

To illustrate each of those four elements and to provide a model that demonstrates how a group of executives and managers identifies its destination and the route they will take, let's observe one of the series of visioning sessions of the executive leadership group of RightCare. RightCare is an integrated delivery system we've invented for the purposes of this discussion. RightCare is a

RIGHTCARE INTEGRATED DELIVERY SYSTEMS

- 4-year-old business entity constructed from 40-year-old components
- Service area: urban and suburban locations (Chicago, Gary, Milwaukee)
- 8 hospitals (50 percent occupancy rate; 2,100 licensed beds)
- 1 nursing home and 1 rehabilitation center
- 2 small skilled-nursing facilities and 1 home-health agency
- 13,000 doctors in IPAs (Independent Practice Associations)
- 12 fully owned groups (350 doctors)
- 8 ambulatory-care centers
- 1.8 million members (HMO)
- $4.5 billion in gross revenues
- 20 percent member growth annually (gross)
- 7.2 million visits annually
- 16 percent revenue growth annually

carefully designed composite that incorporates elements of several actual enterprises: Its core issues are absolutely real. Our entity has "existed" in its current form for only four years, but already it has attained a position many in the industry are striving to achieve. Located in the Midwest, RightCare, a not-for-profit institution, serves both urban and suburban communities, and it is a strong player in the marketplace.

Executives of RightCare hold their visioning session at a suburban conference center that they often use for strategic planning meetings. Here, seated at a round table, the organization's president; the chief medical officer; the chief information officer; the vice president for marketing, sales, and customer service; the vice president for operations; and two member representatives meet with a visioning consultant. All are excited about participating in the creation of RightCare's new vision.

The president stands to define the agenda. She states that the goal of this meeting of the executive leadership group is to set the direction for the next three to five years. She reminds the group that this and other off-site sessions are part of a visioning process that began with the collection of a tremendous amount of information about RightCare's business environment. In preparation for the work sessions, team members spent weeks gathering and studying materials on market trends, innovative health care treatments, competitors' profiles, and customers' needs as projected for the geographical area that RightCare serves. The process will continue with a progressively refined sense of what areas to monitor in upcoming months. She states that visioning is not a once-a-year event: It is the way they will direct RightCare's progress.

THE RIGHTCARE GROUP: DISCUSSING SHIFTS IN INDUSTRY PLAYERS

- The health care industry used to comprise hospitals and private practice physicians. Today, hospitals, physicians, managed-care organizations, and other industry participants are forming alliances, partnerships, and mergers to create virtually integrated organizations.
- Many employers provide and manage wellness and caregiving services in ways that challenge industry definitions. L. L. Bean, Coors, DuPont, Tenneco, and Union Pacific, for example, are developing wellness centers to help their employees stop smoking, control their weight, or learn to exercise more frequently.[1]
- Some insurance companies are forming managed-care facilities, aiming to participate in subscribers' health maintenance and management.[2]
- Aggressive niche providers are entering the health care marketplace, offering services ranging from oncology to long-term care. Outreach clinics for the diagnosis and treatment of diabetes, asthma, hypertension, skin disorders, and so forth, have recently opened both in the U.S. and the U.K.[3]

The Business Landscape

"We're strong," the president asserts, "but we're at a crucial moment in our history and that of the industry. We're here to reexamine our mission and our goals—to reach a shared understanding of the future. We will derive our plan for action from that shared understanding."

The president's "crucial moment" is her concern for how to respond to all of the forces that challenge RightCare: aggressive new competitors, the need to broaden the range of their services, and heightened demands by members and patients, to name only three. After her brief description of those forces, she leads the group into a discussion of the business landscape, the first element we introduced above. What comprises the business landscape, and what makes this a crucial moment? The chief financial officer pipes up. "Large, for-profit chains are entering our market," he declares. "They're gobbling up providers and, with their economies of scale, they can lower their costs. Our per-member-per-month (PMPM) costs, depending on which you examine, are between 5 and 12 percent higher than those of our competitors."

The vice president of marketing, sales, and customer service pushes back his chair and sighs with frustration. "There's no member loyalty anymore," he says. "Members are switching for better prices and what they imagine will be better care. We have an 18 percent defection rate."

"That's because we know more these days," one of the member representatives says. "I use the Internet and other members do, too. We can research our symptoms and illnesses, and we can inform ourselves about everything from cancer studies to treatments for repetitive-strain injuries. Of course that raises our expectations. We expect our doctors to give us better information and respond promptly to our questions. We see new drugs now widely advertised in popular magazines. The network news shows run a story on health at least once a week. Doctors are no longer the only ones who know about diseases and their treatments."

The chief information officer, who, up to this point, has been uncharacteristically silent, clears his throat. "The more members know," he says, "the less in thrall they are to their doctors. It used to be that they saw themselves more or less in servitude to the physician. Now, however, they see themselves as partners with the doctor, not mere patients. That's great, but it also creates a certain loss of trust in—and respect for—care. That's especially true among younger members. That's the reality out there. And as they get their own answers about health care, they put more pressure on us to be as technically fluent as they are."

"We just want things streamlined," says another member representative. "We don't want to have to talk to five people and spend an hour waiting on the phone for an answer we could have found ourselves. Everyone's time is more precious these days."

The discussion continues, flowing freely from one topic to the next. The participants in the visioning process really benefit from the uncensored exchange of ideas and the sharing of perspectives. Their discussion covers a wide range of issues: from features in the business landscape to realities RightCare must evaluate to determine the direction for the future. Most of the issues, like those the participants have already mentioned, are vitally important to players in today's health care industry. Not each issue affects all players equally and new issues continually emerge.

Up until now, the visioning consultant has been prodding RightCare's discussion along as unobtrusively as possible, making sure everyone has had a chance to speak up and probing for more discussion on topics which seem particularly important to RightCare's future direction. Recognizing that the leadership group has covered the most important points and is quickly edging toward exhaustion and repetition, she suggests a break and prepares for the next stage of the session.

Shared Assumptions

In this next phase, the RightCare group deals with the second element of the framework for the visioning session: shared assumptions.

In most planning exercises, executives and managers rarely take the time to explore their assumptions about what they perceive in the business landscape. Most of us act with the unfounded conviction that everyone thinks just like we do when, in fact, that's rarely the case. This is a most unfortunate misapprehension: After all, most people do not base their decisions on the observable facts themselves. They are much more likely to base them on their *assumptions* relating to the facts.

To prepare for the discussion of shared assumptions, the visioning consultant reviews the flip charts posted all around the room. They reflect the group's earlier discussion of the business landscape. She prepares a list of the thirty or so points articulated during the first part of the session. When the meeting reconvenes, she hands each person a copy of the list. "These are the key points we covered in our discussion of the business landscape," she states. "Take a few minutes and go over it. Some of the items on the list are simple facts about the current situation;

THE RIGHTCARE GROUP: DISCUSSING COSTS, REGULATION, AND MANAGED CARE

- Pressure to control costs and stronger emphasis on management of health outcomes are driving the more pervasive use of clinical treatment protocols. Those same influences are contributing to the proliferation of risk-sharing arrangements with partners who might specialize, for example, in the treatment of a particular population segment—say, patients with diabetes, AIDS, or certain types of cancer.

- By the beginning of the next century, the U.S. federal government may take a larger role in health care regulation. So far, though, nobody can predict what that role will be.

- The number and popularity of managed-care plans have grown with phenomenal speed. Approximately two-thirds of the people insured through their employers now belong to managed-care plans. Nine of the ten largest HMOs are for-profit enterprises like Prudential, Cigna, Aetna, and United HealthCare. More than 50 million people—one fifth of the U.S. population—are HMO participants.[4]

others are statements about the future. As you're reading through, I'd like you to think about whether you consider the points inevitable, probable, or merely possible."

She defines her terms: Inevitable means 98 percent certainty that a particular trend or observable fact is going to be a force to reckon with in the foreseeable future; probable means at least a 70 percent certainty; a possible rating means less than

70 percent likelihood that a trend is going to have significant business impact.

Assigning each issue or observation to one of those three categories of likeliness is a matter of intuition, unspecified feelings, beliefs, biases, and emotions. But therein lies its greatest potential. By identifying certain trends as merely "possible" rather than "inevitable," health care managers make their assumptions and opinions public. This helps direct and facilitate the group's discussion and the process of moving toward consensus and focus on the most inevitable of the issues. It's essential for leadership team members—individually and then collectively—to declare their expectations about the future. From those revelations they will create a list of shared beliefs upon which to build their vision of the future. While accuracy in predictions is far from inconsequential, at this stage of enterprise transformation, the goal is not to be 100 percent correct. The point is to air everyone's assumptions so the entire group can play by the same set of rules.

Imagine the dynamics at a board meeting where the chairman—in his own mind—believes a trend to be inevitable, while board members silently expect that the same trend will only possibly continue. If they don't take the important step of identifying trends and understanding one another's beliefs about those trends, the chairman and the board members might start to implement their strategy, not realizing that they are operating under enormously different assumptions.

When the RightCare group members have completed their ratings, a lively discussion ensues. One member representative has just read a book predicting worldwide plagues in the coming century. She views the question of emergent diseases affecting

THE RIGHTCARE GROUP: DISCUSSING DEMOGRAPHICS, TREATMENT ALTERNATIVES, AND WELLNESS ISSUES

- Pressure to reduce costs and the arrival of advanced information technology means more treatment through community- and home-based delivery systems.
- Treatment alternatives—the consequences of therapeutic and economic forces—are gaining currency. Patients and members are giving serious consideration to such nontraditional treatments as acupuncture, massage therapy, homeopathy, and the like, and they want to receive those services from their "traditional" care delivery sources.

health care management as inevitable. The chief medical officer and the rest of the group are divided in their views. The group bats facts, figures, statistics, and speculation back and forth, but the participants cannot reach consensus. The book the member representative read was authoritative and convincing: she won't be swayed. Finally, the president intervenes. "The relevant question to ask is this: How will our beliefs about possible plagues affect our decisions about RightCare's direction and strategies for the next three to five years? What is the magnitude of the impact this would have on RightCare?"

The leadership group can at least agree that they do not need to make immediate plans for such cataclysmic events. Yet, the president, recognizing the potential impact of a new, virulent disease, adds it to the list of trends that RightCare must continuously monitor in its ongoing strategy management process. Although the group is not ignoring the problem, it's no longer on the list of pressing issues that require immediate action. They deal with several other points in similar fashion, clearing away excess baggage and distractions and creating room for matters that require more urgent attention.

The leadership group agrees on the significant business impact RightCare will feel from information technology. They know that a number of technological improvements will revolutionize the management and delivery of health care over the next ten to twenty years. Still, not all technological improvements are created equal. The participants believe that some sort of electronic medical record for patients, for example, will, before long, be the norm. Those records will take the place of today's manila file folders and hospital charts bulging with forms and notes. While no one can clearly see the future, participants in our leadership group know they must prepare for such strongly probable innovations as the electronic medical record.

Continuing their discussion of information technology, the participants agree that another inevitability is that elements of the health care system, both in the U.S. and abroad, will become increasingly interdependent, mainly as a means of responding to consumer demands. They know that RightCare, along with other health care organizations, will need to count on information technology to get to critical information about treatment alternatives, the medical value of certain therapies, the cost effectiveness of others, and so on.

Yet RightCare's information technology delivery is fragmented. A recent assessment of its information technology capabilities revealed that the 13,000 doctors who work with RightCare are only marginally familiar with information technology. The eight hospitals use systems that meet the needs of certain departments, but they aren't integrated across departments, they aren't well-linked with medical technology, and they aren't accessible by RightCare's partners outside the acute-care setting. And, the chief information officer reminds his colleagues, "the patchwork of computers and networks that we have ingeniously devised to keep things running smoothly today are, I am afraid, going to come crashing down as we ask more and more of our systems."

The leadership group turns to another pervasive concern. The members share near unanimous certainty that such critical medical problems as AIDS, Alzheimer's disease, and breast cancer will continue to stress the capabilities of both public and private resources.

"That's true," the chief medical officer states, "and we can't exclude other serious problems like tuberculosis, infant mortality, and sickle-cell anemia."

The vice president of customer service raises another point: the inevitability that certain problems—poverty, drug abuse, violence, unsafe sex, for example—will continue to be "medicalized." Those concerns have become medical rather than social even though their sources are clearly rooted in social conditions. He refers to a study by Leroy L. Schwartz, who argues that the relatively high infant mortality rate in the United States results from the inadequate preparation of very young expectant mothers, especially those in minority populations who live in poor Southern states. The "real problem," according to the study, is not with the health care system, but "upstream of it." In the U.S. alone, approximately one third of the $930-billion health care bill results from social problems, a phenomenon faced by countries around the world.[5]

As the discussion proceeds, the group realizes that they could endlessly debate these problems and their causes. However, when faced with a parent in need, caregivers do not generally ask how one contracted a disease, only how best to deal with it. It was clear to them that chronic diseases and the medicalization of social problems will place increasing demands on RightCare. The medical director intervenes, "We need to turn our attention to how we collectively believe this will affect RightCare and what we must do to anticipate and respond. We need to incorporate these conclusions and build them into our operational vision."

By the end of this part of the session, the group has developed a clear understanding of all those things that will shape their future. The group compiles a list of trends and issues to watch: All the cards are on the table, and everyone is speaking the same language. Even if everyone is not in 100 percent agreement, all members of the group know exactly where the others stand on specific issues. Most important, the leadership team has a common starting point for the vision.

Having spent a good part of the first day debating how changes in the health care environment will affect RightCare, the visioning team decides to conclude that day's discussion. Tomorrow, they will begin to craft an operational vision that responds to what they have seen in their changing environment. It's not enough just to fantasize about the future. Members of the RightCare enterprise need to determine how they will know that they are achieving their goals and how their enterprise will operate in the future.

Operational Vision

When people think about creating a vision they likely imagine crafting a statement that enthusiastically reaffirms their organization's ambition to gain market share, serve customers, and be competitive. Yet, even the most carefully phrased vision statement cannot alone guide the actions of complex enterprises during their transformations.

What organization, after all, doesn't want to be best or to capture the greatest market share? Turning such imprecise ambition for excellence into action requires a more specific set of definitions and goals. The people and the processes in an organization must have clear ideas of where they're headed, how to measure their progress as they near their goal, and how this relates to what is happening or might happen in the market. We call this an operational vision.

When the RightCare team begins its second day of discussion, participants agree that once they have determined the lay of the business landscape, once everyone has discussed and reconciled disagreements on shared assumptions about this landscape, the team can begin to outline an operational vision. Think of this as the road map for the journey of enterprise transformation. The operational vision consists of five key elements: mission, strategic intent, stakeholder value propositions, critical success factors, and performance targets. The team members take a few minutes to define each of these elements:

- *Mission* conveys an understanding of overall purpose of an enterprise and the industry it represents.
- *Strategic intent* guides and inspires action.
- *Stakeholder value propositions* define the unique ways the business will contribute value to each of its stakeholders.
- *Critical success factors* are those achievements that will define success. To make the operational vision a reality, the organization must accomplish each of those factors extremely well.
- *Performance targets* set expectations for and define indicators that tell the organization when it's successful.

At the second day's morning session, the visioning consultant for RightCare explains that strategic intent captures the essence of winning and provides consistency to short-term action while leaving room for reinterpretation as new opportunities emerge. It sets a target that deserves personal effort and commitment. Strategic intent envisions a desired leadership position and establishes the criteria the organization will use to chart its progress.[6]

It also encompasses an active management process that focuses the organization's attention on the essence of winning, motivates people by communicating the value of the target, leaves room for individual and team contributions, sustains enthusiasm by providing new operational definitions as circumstances change, and consistently uses intent to guide resource allocations. In short, it describes a competitive end point for an organization, proclaiming what it intends to do (not how it will do it), and creating an image around which those directly involved in the transformation effort can rally.

The Japanese manufacturer Komatsu set its strategic intent, "Encircle Caterpillar," which employees found much more inspiring than "Be the world's leading manufacturer of construction equipment." At the time, Komatsu was less than 35 percent the size of Caterpillar. Canon cried "Beat Xerox"; Coca-Cola intends to "put a Coke within 'arm's reach' of every consumer in the world."[7]

Citing those examples, the RightCare visioning consultant summarizes the team's analysis from the first day and synthesizes the information into a preliminary draft of the enterprise's operational vision. Although the participants know that they must refine the entries, they believe that they share a sense of common goals and objectives.

The group members engage in a lively debate about the draft statement. As they discuss and refine each point, they realize that they must come to grips with their own organizational strengths and limitations: RightCare's "core competencies." Core competencies are the essential skills that differentiate a company from its competitors.

That quickly leads everyone to consider each element of the operational vision and the effect it will have on RightCare's business processes. For example, to set the standard of success

RIGHTCARE'S PRELIMINARY OPERATIONAL VISION

Mission

- To provide quality health care through an integrated system of health care delivery and financing.

Strategic Intent

- To break all the rules and lead the industry in delivering *truly* integrated health care.

Stockholder Value Propositions

- To promote wellness for members, patients, and the community. When care is needed, it will be personal and excellent.
- To commit to the achievement of mutually beneficial objectives—all in the service of excellence in health care—for our providers and business partners.
- To foster an environment that values initiative and learning and provides opportunities for RightCare's staff to achieve personal goals.
- To offer efficient and cost-effective services for purchasers.

Critical Success Factors

- To provide—through integration with alliance partners—seamless care delivery services to members and patients across the continuum of care.
- To set an exemplary standard of service for purchasers, members, and partners.
- To improve overall patient health and strive for the best achievable health care outcomes.
- To reduce cost per-member-per-month to the lowest level of all comparable integrated delivery systems.
- To know our members, no matter what facility they use or what services they choose.
- To continue to grow, steadily increasing our market share as well as our geographic coverage.
- To create incentives for providers, partners, and employees, which directly support and encourage actions in line with RightCare's mission, vision, and values.

Performance Targets

- To achieve member surveys that rate continuity and quality of care as "exemplary" 98 percent of the time and have members report enthusiastically that the service and quality they received was beyond their expectations.
- To avoid losing members because of inadequate service.
- To retain members as one of the top 10 percent of all integrated delivery systems in the nation.
- To be recognized in professional publications as a model for the achievement and management of health outcomes.
- To have stakeholders—our employees and partners—rate RightCare as "the very best overall" 95 percent of the time.

for service, the group starts by discussing all the parts of RightCare to determine the contribution of each of those parts to the achievement of that success factor. Participants understand that in addition to the customer service process, triage and advice activities, scheduling and referrals, and contracting with alliance partners, virtually every aspect of RightCare's business will need to be involved. Everyone is beginning to see just how different this visioning approach is from earlier forays into "strategic planning."

As they proceed, the visioning consultant's admonishment that "words matter" becomes clearer. The discussion of the critical success factor related to growth reveals that some members of the group interpret that growth as increasing membership, whereas others perceive growth as expansion of geographic reach. Before long, the members of the group recognize the folly of proceeding without first defining terms. They form smaller groups to accomplish that task, refining their statements so they are precise, unambiguous, and specific. This clarity is important so that the final vision can truly guide the actions of others. To be sure, the matter of bringing the entire organization on board and enlisting the assistance of all employees remains no small matter, but it is far from an impossibility.

Given the choice of languishing by the wayside or earnestly undertaking the journey to enterprise transformation, most health care workers do rally to their leader's side. But leaders must first enlist their participation and enthusiasm. Outside the realm of health care, we find similar indications of increased employee commitment: For example, the advertisements for United Airlines present a company owned by the people who work in it. As Hamel and Prahalad point out,[8] it is difficult to imagine workers rallying around a company's strategy merely to raise the stock price and further enhance shareholder wealth. A strong leader with an expansive operational vision knows the important roles that all people in the organization can play, perhaps even in ways that they themselves don't yet recognize.

When Bill Gates addressed a group of financial analysts recently, he announced that Microsoft was "hard core" about its intent to seize the Internet market from current leaders. The company, he said, was prepared to devote all its energies to that end.[9, 10] Saying and doing are, however, two different things. Gates' success indicates that passion for "doing" brings results.

While some managers are content to plan next year's budget or reallocate resources in support of a new care path, others are tearing down walls—both real and figurative—between departments, and they're adding information technologies that will prepare them for major shifts in the health care industry in the coming decade. Managers encourage action by creating a sense of emergency or crisis. Every operational vision should be infused with such urgency and with Bill Gates' kind of passion. As the members of RightCare's leadership group and all workers in health care know, there is a crisis in the industry today.

Extraordinary times call for extraordinary actions. As we saw in the case of El Dorado, Sandra Smith's proposal to cut expenses by 50 percent left the organization's managers gasping for breath, but once they found their feet, they set out to create a wholly new organizational structure to replace the multi-leveled managerial hierarchy. As we noted in Chapter 2, Smith's announcement was not intended to initiate cost-cutting; it was, however, designed to fire up El Dorado's engines and put its people on the fast track to victory.

RightCare's leadership group has its work cut out, starting with listening to people throughout the company as they are exposed to the emergent vision. Top leaders must persuade everyone connected with the organization to take a personal interest in exceeding the competition's best efforts. In addition, employees must learn to be good team players and to solve problems creatively and collaboratively. As the assessment of RightCare's current information technology environment concluded, the overall quality of RightCare's information technology platform needs considerable overhaul. To perform properly, everyone in the organization must have access to the best available technology. It does little good to help employees develop their skills as members of a care path team if inadequate information systems frustrate their efforts to communicate and work efficiently.

The Course of Action

"We've come a long way," RightCare's president announces. "I think each of us has a better understanding of where we are and where we're headed. We know what we're trying to achieve and why, and we have set measurements for success along the way. Our last task is to determine the right course of action."

As a concept, the course of action is perhaps the simplest and most self-explanatory of the elements we've laid out. Putting it into practice, however, is often the most difficult.

RightCare's leadership group begins by reviewing the experiences of other organizations that have undertaken the kind of change we envision, debating the merits and pitfalls of the journey each followed. No two companies can travel identical paths, and only the most shortsighted resist learning from the mistakes and triumphs of colleagues and competitors.

Armed with competitive intelligence and the operational vision, the group itemizes the accomplishments that will make their vision a reality. They review business processes, how people work and what motivates them, and how to make the most of information technology. They discuss alternative ways of tackling their objectives and begin to make choices about the most promising ways to achieve their vision.

The discussion, which the visioning consultant has been recording on flip charts around the room, yields, when analyzed and aggregated, a detailed list of twenty-five "change programs," each of which will ultimately consist of several projects. Approximately half of those programs fall under the heading of business actions—encompassing changes to processes, organization, people, and business strategies—and half are in the crucial information technology category. When the group looks at the list, a wary silence falls over the group.

"Is everyone feeling as daunted by this as I am?" the president asks.

"It's unlikely that you'll be able to undertake and absorb so many projects and programs: In fact, it's probably unwise even to try," the consultant says. "We have one final step. We need to go through the list and pick out the three change programs you would implement if you could implement only three."

"*Only* three?" everyone asks in unison.

The consultant assures them that what they are about to do

RIGHTCARE'S PRIORITY CHANGE PROGRAMS

- Eliminate redundancy and increase use of specialized resources by *coordinating care delivery across all components* of RightCare—including ambulatory services, acute care, emergency services, primary care services, and home health.
- *Create a member-service capability* that enables caregivers to resolve clinical and administrative problems without frustrating or inconveniencing members. Ask our members to describe how they want to be served and reengineer the customer service process according to members' preferences for treatment and administration. Combine this with a twenty-four-hour hotline for nursing advice and triage.
- Create the capability to *provide clinical information* to all personnel responsible for decisions affecting individual patients at the point of service and the time of care delivery. A key element of this will be an information system for clinicians and providers that will let them get to a patient's records at any time, from any place, including remote locations.
- Institute a broad-based *preventive-care program* for our members and patients, to include proactive follow-up, wellness and prevention classes, reminders at regular intervals, and so forth.
- Provide *medical information* in ways that capture the imagination and attract new members: on-line dialogue, question-and-answer services, and Internet connections. Educate those members who don't yet know how to use information technology as their health care partner.
- *Improve service* in order to retain members while *decreasing costs*. Establish incentives that attract "highly desirable" members.
- Create an *information technology infrastructure* that integrates the enterprise end-to-end. In particular, it must permit all facilities to share and use lifelong medical information about individual patients. It must also enable IPA physicians and other partners in the RightCare family to communicate with one another and with RightCare.
- Develop a program—including the use of Care Paths, tracking services, and encounters across the continuum of care—to achieve *exceptional health care outcomes*. Report information about outcomes to achieve a competitive advantage. Educate everyone in the organization about the processes and technologies that will enable RightCare to attain the best medical outcomes in the industry.

is to set priorities. As lively as their original discussion of assumptions regarding the business landscape had been, their debate about which programs should take priority really takes off. Everyone is, however, amazed by how much consensus there is among the group members. People agree about the need to integrate and coordinate care across all aspects of Right-Care's enterprise, to eliminate redundant resources and to improve customer service. In addition, they realize that they will need to share member and patient health records through-out RightCare and with all 13,000 of their IPA physicians as well as with other care-delivery organizations who see Right-Care patients. They produce an amended list: the final blue-print for RightCare's action.

"Congratulations, ladies and gentlemen," the president announces. "We have an action plan in hand—a place to begin. Let's move forward. We've made a great start."

Postscript: A Memo from RightCare's CEO to All Employees

I am delighted to be able to give you an update on the results of the work we have been doing to create a new vision for RightCare.

As I have shared with you in previous communications, we tended to look at the future in terms of business planning. This year, as we thought about becoming the truly integrated delivery system we all want, we realized that it was essential to step back and take stock of the business environment we operate in and of our own strengths and weaknesses within that environment.

We have immersed ourselves in what's going on in the entire industry and in our own local market. We have looked at the trends in payment for health care, disease management, outcome management, and consumers' needs. And we have looked very carefully at what RightCare excels at and what we need to improve.

Armed with that information, we stepped back and asked ourselves: What effect do we think those trends will have on us? What responses do we need to make if RightCare is to match our ideal?

Frankly, we didn't like some of the answers to those questions. We discovered that in certain areas, we weren't as strong as we ought to be. To compensate for some of those shortcomings, over the next few months, we will explore a variety of complementary mergers and alliances with other organizations.

Those activities represent only one aspect of the detailed course of action we have set for RightCare. We have called for change throughout RightCare: in how we operate, work together and with other health care organizations, and integrate information technology.

Let me give you an example. One of these programs for change addresses the challenge of resolving members' clinical and administrative issues in a one-stop shopping mode. To achieve that goal, we will install new information systems that tie together our data about patients, providers, the care provided, alternatives for clinical interventions, wellness programs, administrative matters, and so forth. We will redefine our processes for providing service to our members, starting with asking them what they value. And we will begin an extensive program that will prepare our customer service staff to operate in this new environment.

That change and many others like it will test our ability as an organization to come to grips with the challenge of making information technology a real partner. Each of us will have to be more capable in our use of information technology!

As we progress, we will all make changes in how we work. We will start thinking of technology as our partner, and we will find the best ways to work with that partner to achieve excellent results.

Clearly, we want to be in the vanguard of health care as we move into the future. We can achieve that only by keeping clear sight of our vision and knowing what we want to become: the best possible integrated delivery system in this region.

We need your energy and your enthusiasm as we all build a common vision of RightCare's future. I know that each of you is committed to the long-term success of RightCare. I hope that each of you is excited about where we are headed. And I know that each of you will benefit individually as we make RightCare a continuing success. In my next letter, I will announce a series of face-to-face meetings at which we will begin to share and discuss RightCare's new vision.

Thank you for your continued good work with RightCare.

CHAPTER 4

IN THE INFOCOSM OR OUT OF BUSINESS

Perhaps we can all be forgiven for not fully appreciating the firestorm of change sparked by information technology. After all, not many people in the late eighteenth century foresaw the radical changes that the Industrial Revolution brought to civilization. By their very nature, such radical changes transform the landscape in ways that few people can visualize. Today, many of the old business operating assumptions are crumbling. Altogether new kinds of organizational assets—speed, agility, reach, and insight—are rising up in their place.

We don't need a clairvoyant to see what changes are possible—we would say inevitable—in the health care industry as it comes to grips with the power of information technology. Certain effects of information technology are already clear. We know that in the 1970s information technology played a support role in the business, making such routine functions as accounting and order processing more efficient. In the 1980s, the advent of personal computers and networks put computing in the hands of users and they began to change the way they did business: analyzing customer service patterns, consumer preferences analyses, and marketing trends, and modeling the design of new products and services. In the 1990s, we are realizing our earlier investments in end-user technologies. We have the ability to move decision-making closer to the customer and create essential knowledge at each step in business processes.

The most tangible evidence of this change appears in how health care enterprises define and execute their business processes. Indeed, how a health care organization conceives and defines its business processes is a direct reflection of how it defines its business. For example, an organization that sees its fundamental purpose to be keeping *members well* rather than treating them once they become sick will place significant emphasis on such wellness-related processes as the analyses of member demographics and needs, the development and dissemination of education and self-care information, and the execution of care coordination. Conversely, an organization that sees its main mission as effectively and efficiently treating *illness in patients* will focus much more on processes such as the management and delivery of primary and specialty care.

In the Health Care Infocosm, clear operational visions and the process choices that follow from them are increasingly vital as organizations are able to put information technology to uses that were not possible only a short time ago. An effort to envision this future was recently undertaken by a group of multidisciplinary leaders of the Australian health care industry who constitute the Health Futures Forum.

This group considered the factors influencing the industry and created four possible future scenarios (Figure 4-1). They concluded that the future health care world they aim to achieve in their country must be "integrated and virtual." In this scenario, the predominant health care model will integrate *virtual* organizations through information technology to create the Health Care Infocosm. Furthermore, there will be an enormous increase in the number and extent of global alliances. These

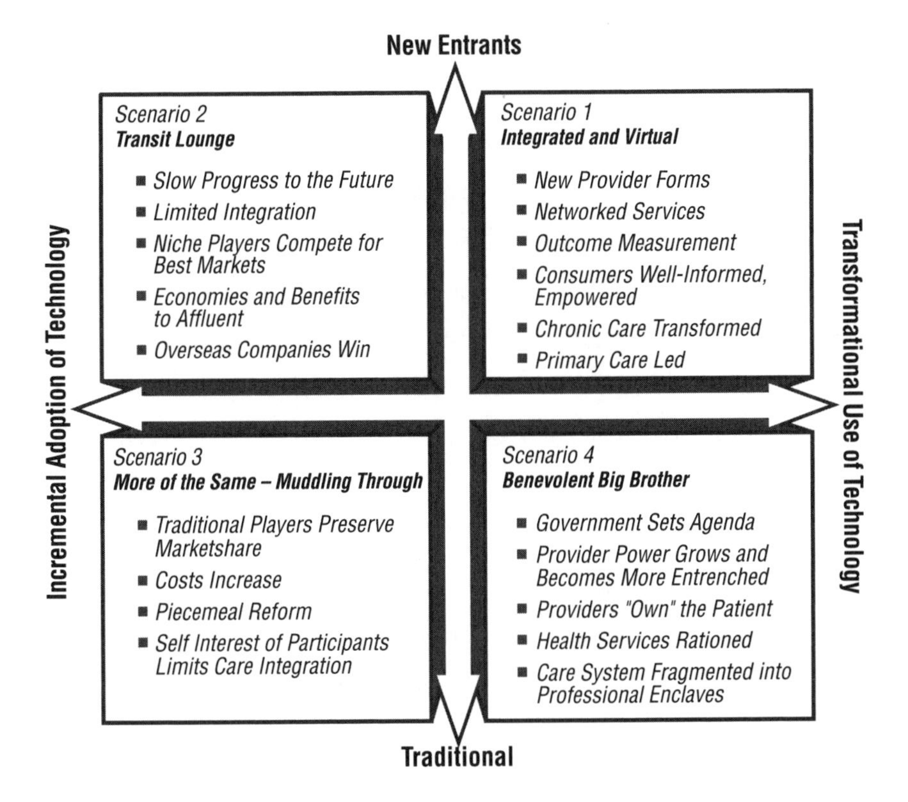

Figure 4-1. Australian Scenarios for the Future of Health Care. (Source: Oliver, Morlet, and Weerasingham 1995.)

---

health care leaders imagine seeing "strange bedfellows," even those who were once fierce competitors, cooperate and collaborate in ways that are mutually beneficial.[1]

This scenario epitomizes the way that combining and recombining business processes will occur in the Health Care Infocosm. The success of these virtual organizations, however, rests squarely on the sophisticated and effective use of information technology.

After all, it is within business processes that new strategies enabled by new technologies come together to create new assets. Using the power of telecommunications, for example, "plug-and-play" business processes can be woven together more creatively than before, yielding enhanced efficiency, reduced cost, and strengthened core capabilities. Ultimately, the most successful organizations will be virtual ones, multiple entities linked together through shared interests and coordinated services in such a way that renders the boundaries between them invisible.

RightCare, which we introduced in Chapter 3, recognized that only by dramatically revamping its business could it successfully compete in the future. RightCare's executives saw that to provide the cradle-to-grave services they envisioned, they would have to make substantial changes—not just inside RightCare but beyond its walls—to processes shared and executed in concert with other organizations either owned by or affiliated with RightCare.

This shift beyond an organization's walls first demands a process orientation that begins with building skywalks between what may well be monolithic departments or functions. As illustrated in Figure 4-2, the adoption of a process orientation and its ultimate utility in making cross-enterprise competencies possible usually take place in three stages.

In the first stage, organizations acknowledge the feasibility of processes, yet continue to allow functions to dominate their business. As we will see in Chapter 5, people tend to hold dearly to old concepts of working in functions where clearly defined responsibilities and rewards are possible.

In the second stage, processes emerge as the key drivers of business activity, and the organization looks very different from

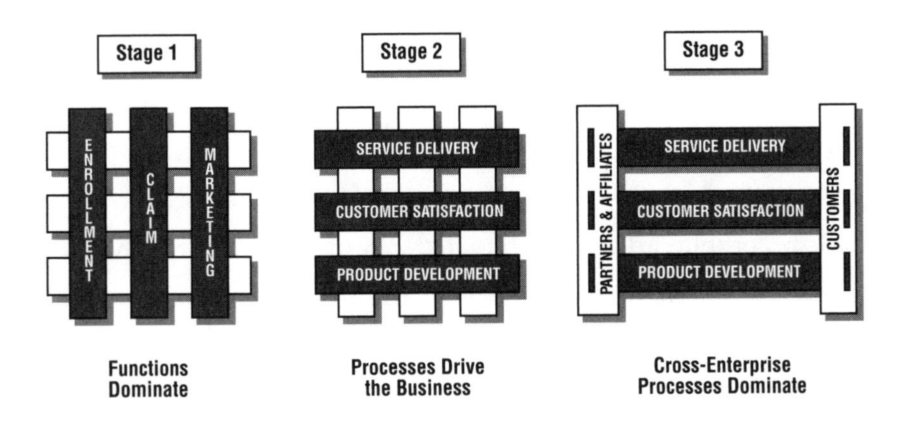

Figure 4-2. The Transition to Cross-Enterprise Processes.

earlier incarnations. Process skywalks begin to be constructed between functions, allowing the free flow of people and ideas. "Customer satisfaction," for example, encompasses activities oriented to attracting new members, enrollment, and following up on claims.

In the third stage, processes actually extend across and among organizations, improving quality and service as well as reducing costs. This extension supports a true virtual organization, where interorganizational linkages are the norm and where "mass customization" of products and services for individual members and customers is enabled.

Astra Merck Inc. is a model of the power of processes to create wholly new ways of doing business. Astra Merck, a billion-dollar pharmaceutical enterprise headquartered in Wayne, Pennsylvania, was born out of a joint venture between Merck & Co., Inc. of Whitehouse Station, New Jersey, and Astra AB of Sweden. Astra Merck began with an ambitious goal: Reinvent the process of selling drugs. The changing environment of the health care industry is the opportunity Astra Merck intended to exploit, by "wrapping their drugs in free information services," and billing themselves as an enlightened "pharmaceutical solutions partner."[2]

One particularly important benefit of the shift to a process orientation is the focus it directs to the customer. People working in processes interact directly with other teams, suppliers, and stockholders, but, most important, they work directly *with customers* rather than *for bosses.* Nancy J. Wysenski, executive director of field sales at Astra Merck, succinctly puts it this way: "The difference in a process organization is that it forces us to stay focused on the customer as opposed to a typical organizational design where you become internally focused."[3]

Every health care enterprise has many customers: members and patients, regulators, business partners and affiliates, providers, purchasers, and community organizations, to name a few. Because each set of customers puts different demands on the organization, a clear definition of who the customer is and a thorough understanding of what that customer needs become essential elements in the successful transition from functional to process organization.

The inevitable transition to a process orientation yields benefits of efficiency and customer focus. The balance of this chapter will discuss two other imperatives for industry leaders:

- To extend processes beyond their own organizational boundaries, and
- To conceive and imbue these processes with the full range of technology's powerful reach.

Extending Competencies with Cross-Organization Processes

Successful leaders must extend core competencies by forging relationships across organizational boundaries. As medical technology and information technology have redefined health care on a global scale, no single enterprise can provide the complete range of health care support that members will demand.

Health care organizations make substantial investments of time, energy, and resources to provide services across the continuum of care, in which wellness and care services are managed seamlessly regardless of where services are provided—in outpatient facilities, in acute-care settings, in rehabilitation settings, and so forth. Nearly always these extensions require organizations to reach beyond their boundaries for competencies and capabilities they don't possess and that others do, such as patient education programs, preventive medicine and health screenings, rehabilitation medicine, or urgent care.

Many health care organizations are eager to form partnerships and alliances in order to expand their geographical reach and to reduce their operating costs. Distinguishing between mere duplication and complementarity, these new alliances recognize the imperative to create and foster cross-organizational processes. *The Economist* recently noted that in 1994, the health care industry saw 1,100 acquisitions and mergers worth approximately $60 billion.[4]

Health care organizations are combining to avoid duplicating services and to give members access to a broader range of services. In January 1996, for example, three major health care institutions in Boston—the Dana-Farber Cancer Institute, Brigham and Women's, and Massachusetts General Hospital—announced the creation of a network called Dana-Farber/Partners Cancer Care, Inc. While not a corporate merger, the network combines the clinical services of the three health care institutions: The three facilities can share the benefits not only of more convenient access to health care but also of cross-organizational processes for improving care delivery. Executives estimate that the network will treat some 15,000 adults and will handle about a quarter of a million outpatient visits each year.[5]

The cross-enterprise processes that are created through such partnerships are nothing if they do not provide accuracy, cost efficiency, and seamless care delivery. Customers must see no rough edges, no tears in the fabric of the virtual health care enterprise. At the same time, partners in the relationships must establish ways of measuring and sharing information about economic and clinical results. They must establish consistent standards for wellness and care and for customer service, so to achieve the promise of seamlessness.

They need to establish ways of managing staff who "belong" to one organization but who are indistinguishable from another's in day-to-day operations. Also, they must establish how to deal with the organizational interdependence that grows with the effectiveness of cross-organizational processes. We will examine this issue more fully in Chapter 5, where we take up the need to recognize and reward people for their insights and innovative solutions. Perhaps the major lesson here is that knowledge must not and cannot be restricted by the artificial boundaries of any organization.

Ken Murtha, who participated in the design of Astra Merck and now directs its Product and Customer Operations process, has a particularly informative vantage point for viewing processes: "The process organization demands a complex, interwoven group of skills that you may not have had to manage before. This means adopting a management mode with lots of different permutations scattered throughout the processes. That's the big issue."[6]

Imbue Processes with Information Technology

Health care organizations must incorporate information technology and its benefits into their process design to realize their visions. This imperative meets immediate challenges because today's information systems often support yesterday's out-of-date functions. Even before they shift to a process-oriented view, many organizations find that their ability to manage the human and technological resources required to support their organizations is out of date.

Beyond problems with deficient hardware and software, organizations find other problems: Relationships are too varied, too complex, and often too far removed. For example, a managed-care organization that contracts with one company for wellness and member education services, another for mental health, another for orthopedic rehabilitation, and a fourth for payroll processing is either in the Infocosm or it's out of business. Increasingly, we see this pattern where sets of institutions are linking their processes together in an interdependence that provides unprecedented customer value.

It would be impossible to operate in those cross-organization models without information technology. First and foremost, information technology makes it possible to share services and member information. Already we can see this scenario being played out. The Sisters of Providence Health System in Portland, Oregon, has two 400-bed hospitals, three smaller acute-care facilities, an 80,000-member HMO, a 120,000-member PPO, a primary care division, and a home services unit. The parent organization created an electronic medical record system that is available to all facilities through a sophisticated metropolitan network. Its physicians are well positioned to manage members in multiple clinical settings, including critical care, medical/surgical care, primary care, and home care.

Information technology can make routine work of resource and knowledge sharing among processes. For example, the National Marrow Donor Program tracks and matches volunteer donors with patients who have blood-related diseases. Its system uses telecommunications to link a nationwide network of donor centers, transplant centers, bone marrow collection centers, and testing laboratories.

In another example of knowledge sharing, Astra Merck established a Lotus Notes© network that lets salespeople working with customers use their laptops to record—and make available to colleagues—details of managed-care plans and procedures, contracts, drug studies, and other pertinent information. When the company started to sell Prilosec®, a drug used for acid-related gastrointestinal disease, it collected ideas from around the country about which sales methods, informational and educational approaches, and service strategies worked best and which failed.[7] Few competitors have done the same.

Those examples suggest the critical role of information technology in successful process management—within and across organizational boundaries. However compelling, they represent only single uses of information technology. As Nancy Wysenski of Astra Merck knows, this is not enough:

> All along, we put pieces of the technology puzzle together. Then, on top of that, we superimposed all the communication tools—Lotus Notes, E-mail, voice mail, videoconferencing. It all just rolls together into one huge, effective package. It is the synergy of all the pieces that come together to make you successful.[8]

In the Health Care Infocosm, cross-enterprise processes will be more tightly linked while corporate borders will blur. People, time, and resources are all freed for what counts most: delivering value to customers.

An Image of the Future: Technology-Enabled Processes

Speaking recently of the changes in computer technology, Bill Gates said:

> We are watching something historic happen, and it will affect the world seismically, the same way the scientific method, the invention

> of printing, and the arrival of the Industrial Age did. Big changes used to take generations or centuries. This one won't happen overnight, but it will move much faster.[9]

What will this Health Care Infocosm really look like? How will enhanced value truly be created?

This section describes how a number of the processes might look in the future. We introduce an example of a process model for an integrated health care delivery system (see Figure 4-3). This high-level process view will share common features with many health care enterprises, most of which will increasingly find themselves operating as components in an integrated delivery system.

The model reflects the key objectives of providing high-quality wellness and care services to members and creating, marketing, and managing innovative products. More specifically, the processes center on three areas: first, Health Care and Clinical Capability, which focuses on keeping people well, managing their care, and building care delivery capabilities; second, Marketing, Sales, and Product Management, which deals with crafting, selling, and managing innovative products and services; and third, Support, or those processes that provide a strong foundation to make all of the other processes possible.

Let us take a closer look at three of the processes in this model. First, we examine "Provide Wellness and Care Services" and several of its subprocesses. Second, we'll look at "Manage Delivery System Capability." Last, we'll explore the ways information technology plays a part in "Manage Customer Relations," one of the support processes in this model.

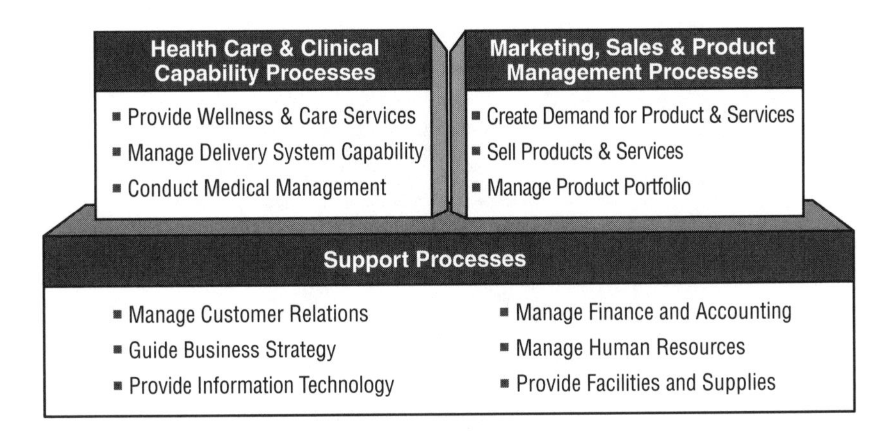

Figure 4-3. A Process Model for an Integrated Health Care Delivery System.

Of course, it is the flexibility, nimbleness, insight, and risk-taking behaviors of organizations that will lead them to achieve competitive differentiation. And those organizational attributes—coupled with the ability to form and manage cross-organizational processes and to exploit information technology within processes—will separate the winners from the losers, the innovators from the laggards.

Provide Wellness and Care Services. Perhaps the possibilities of the Health Care Infocosm will be most visible in the Provide Wellness and Care Services process. This process covers a broad set of subprocesses, from identifying opportunities for member education, self-care, and wellness programs, to conducting triage and providing member advice, to care coordination and case management, to the delivery of primary and specialty care itself (see Figure 4-4).

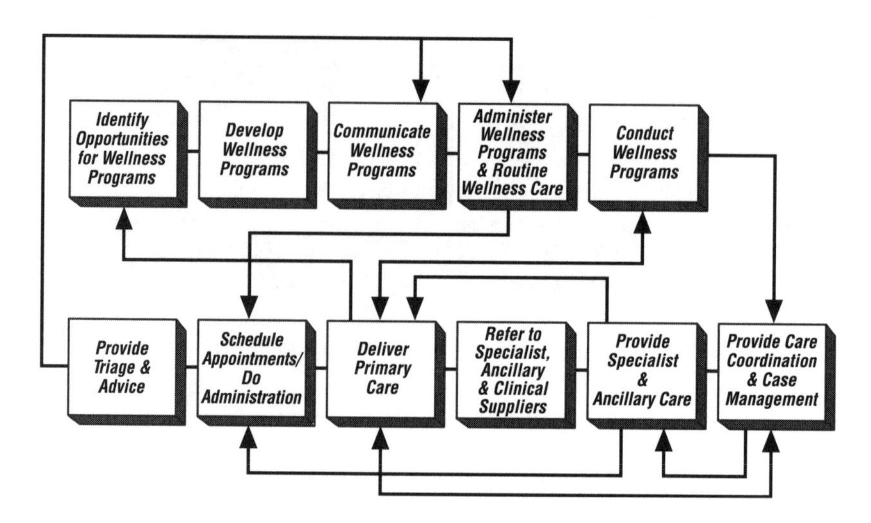

Figure 4-4. Subprocesses Comprising the *Provide Wellness and Care Services* Process.

Wellness. The wellness subprocess strives to prevent acute health care problems. Keeping people well is not only good medicine but good business. Indeed, promoting wellness effectively is likely to become a survival criterion for health care organizations in the coming years.

Today, in the world's wealthiest countries, people over age 65 consume about one third of the total amount spent on health care. If current trends continue, by the year 2000 their care will account for half of that total.[10] By promoting wellness, health care organizations can manage the proportion of the care demanded by this growing population segment while it simultaneously provides the benefits of improved health to all.

The key to wellness is education about lifestyle choices. We know that certain behaviors—smoking, alcohol and drug abuse, obesity and unsafe sex, for example—account for approximately

40 percent of treatment costs for what could otherwise be largely preventable medical conditions such as lung cancer, AIDS, and tuberculosis.[11,12]

Still, few health care organizations include wellness programs on a list of their core competencies. They depend on relationships with organizations specializing in wellness service to meet members' needs. Not only must wellness services be provided seamlessly to members, but also information about them must become part of their medical records regardless of whether a health care organization provides the care or relies on a partner for doing so.

Effective wellness programs depend on information technology to make them valuable to citizen-patients. For example, it can create on-line access to information about symptoms, diseases, treatments, and outcomes. Already the Internet offers twenty-four-hour-a-day disease-specific discussion groups like those for AIDS or cancer patients. The Columbia University Health Service offers anonymous, on-line medical consultations through its "Go Ask Alice" site on the World Wide Web.[13] The prevalence of home computers will allow wellness and care-related information to be beamed into our living rooms and home offices.

A more focused derivative of this member insight might be dubbed private education. In this way, we can focus advice for such target groups as those who may be at high risk for coronary artery disease. Scanning an entire patient population, information technology identifies people in the risk group and initiates focused information flow that encourages appropriate preventative activities.

In both of those cases, information about members' involvement in wellness activities becomes a part of their overall medical records. As an additional benefit, life care plans can guide, track, and record interventions as a way of helping define and understand care delivery outcomes.

This infusion of information technology altogether changes the wellness subprocess and allows us to realize tremendous efficiencies. Now the critical criterion is the prompt availability of information whenever a member or patient wants it. In addition, a technology-enabled wellness process provides the basis for truly understanding the impact of wellness activities on long-term health: We can track the impact of information and wellness activities on peoples' behaviors.

Triage and Advice. Triage and advice are other key elements of care delivery. Growing demands for round-the-clock customer service require emergency as well as routine assessment and advice, often by telephone. As such, this subprocess becomes the major strand of a giant web of information and resources. It must provide immediate access to knowledge about prevention, diseases, and emergency treatments; about resources that are available; when and where those resources can be accessed; and about interventions and outcomes. Because the "advice nurse" never knows the nature of the phone call, the information needs of such a service are enormous.

The triage and advice subprocess moves into peoples' homes and workplaces when they use any of the numerous software packages currently available for home computer use. For example, Medical HouseCall™ and Iliad™ provide detailed information about thousands of health-related conditions, databases for medical hotlines and information services, as well as cost analyses for treatments and drug therapies in various geographical areas. Equipped with this new knowledge, people are more likely to participate actively in their own health care and wellness programs.

Another critical part of the information technology needed to enable triage and advice is information on the type and nature of care delivery resources available. With full knowledge of schedules, hours, or other restrictions on available services, the advice-giver can not only make a judgment about what is needed but also direct a member to the proper location.

The triage and advice subprocess, as described here, is transformed by information technology. No longer is the advice limited by the knowledge of the person answering the phone. When they are linked to the right information, members themselves can play a role in initial triage. Questions are answered immediately, facilitated by access to a broad range of information on diseases and treatments.

Care Coordination and Case Management. Far beyond the traditional "utilization management," this subprocess is essential to the smooth integration of health care resources and patient and family involvement. Organizations such as SelectCare, a mixed-model health maintenance network in the Detroit area, recognize the power of information technology to enable HMO members and their families to become more fully engaged in disease prevention, self-diagnosis, and the implications of lifestyle choices.[14]

Like the other subprocesses, care coordination and case management are transformed through information technology. At the core of this coordinated effort, a life care plan defines standard approaches to wellness planning and care delivery for particular conditions—everything from childhood asthma to breast cancer. Using life care plans, health care personnel can aggregate members' experiences to create an important source of information for researchers and health care professionals.

Drawing on information from these many sources, caregivers can better coordinate and manage the health of their members. A comprehensive medical record will allow health care givers to identify any situation for monitoring and enable them to provide support for it. It allows them to track the progress of care and to advise the family and the member of expected milestones and areas where they may need to become more involved. In fact, in the early days of care paths, Alliant Health System in Louisville, Kentucky, found that without comprehensive information technology, people had to compile reams of completed care paths and enter each record into a PC system.[15]

Care Delivery. The care delivery subprocess itself is, perhaps, the one most transformed by the promise of the Health Care Infocosm. Here is where the vast health care enterprise touches the citizen-patient. The art and science of medicine come together to keep people well and to examine, diagnose, prescribe, treat, and monitor them when they get sick. The lifeblood of effective care delivery is the information that will tie all of these activities together, wherever and whenever they occur across the continuum of care. "To stay ahead of the industry transformation, leading organizations must improve the way they manage information. This requires a comprehensive understanding of the information needed and the interdependencies of different types of information. Improving access and use of information is a critical success factor in keeping pace with—or being a leader in—the industry's transformation."[16]

Electronic medical records for each individual will be among the significant benefits of information technology. Every one of those records will contain a complete demographic profile and family history for each citizen-patient, as well as a view of wellness activities, care, and outcomes over a lifetime. To understand patterns of symptoms, diseases, treatments, and results, caregivers will aggregate information on huge numbers of citizen-patients. That will be invaluable in assessing, for example, which treatments or drug therapies achieved the best results at the lowest costs.

The electronic medical record will include much more information than today's medical records. It will include health care information in many forms: text, numbers, video, audio, graphs, x-rays, DNA maps, and a host of others. The electronic medical record, which can contain thousands of pieces of information, will be a vast storehouse of observations, diagnoses, treatments, and prescriptions. The record will rely on a free-floating, non-centralized record system, part of the Health Care Infocosm that will permit authorized users unrestricted access to medical records while keeping unwanted users away. Firewalls—electronic boundaries that restrict access to specific sections of a network—will help control the spread of sensitive information.

Suppose, for instance, that a physician knew that a woman had a genetic predisposition to breast cancer. If the doctor wanted to compare the woman's medical profile to those of thousands of other women who shared such characteristics as age, geographical location, degree of stress, diet, and behaviors, the physician could, in an instant, review 10,000 medical records and extract a set of comparable patients. Such access to health care data will provide medical knowledge that has never before been available.

Another way in which information technology will redefine care delivery is by supporting the consistent use of life care plans. Using care plans calls for a triad of new capabilities: the electronic medical record, a repository of medically agreed-upon care plans covering specific conditions and diseases, and a mechanism for aggregating and analyzing information relative to actual experiences.

A simple example can show how this triad works. An elderly man comes to an acute-care facility for hip replacement surgery. The orthopedic surgeons in the facility have agreed on a generic care plan for hip replacement. When his surgeon reviews that plan and determines there is no need to modify it, she inserts it as a guide in the inpatient record. The care plan forecasts the actions of the man's care team from pre-op, through surgery, recovery, and rehabilitation. The exact details of the patient's course of treatment are recorded; and, in this case, the team notes only a few minor deviations from the care path. When the patient leaves, his actual experience is added to that of everyone else who has had hip replacements. Referring to the growing repository of experience, orthopedic surgeons are able to reassess from time to time the adequacy of the hip replacement care plan. The results? Streamlined care; systematic capture of the care delivery experience; resource-planning information for the care team; and care support services.

In the Infocosm, care will be further transformed through various telemedicine capabilities that let care teams seek or provide advice and assistance around the world—without the constraint of time and place that exists today.

All of those advantages depend, of course, on sophisticated information technology capable of carrying information from site to site, even to far-flung political or professional empires. As citizen-patients move from city to city and change health plans, their records move with them. Even in their initial visits to an emergency room or an internist, their records are available to a care team trying to understand the possible cause of their symptoms.

Manage Delivery System Capability. This process shapes and maintains the clinical capabilities of an integrated delivery system—the wellness and clinical resources and capacity—through which the organization delivers products and services. It includes both detailed data analyses for planning for delivery needs to providing feedback to clinicians and other clinical resources on their performance against the organization's medical management agenda (see Figure 4-5).

This process sets the strategy for where and how services will be provided in both new and existing markets—whether through owned or affiliated resources. It develops a thorough understanding of external market conditions and trends—obtained from research and analysis from many sources, including clinicians, the sales force, and purchasers and members—as well as information about the delivery of services from within the organization.

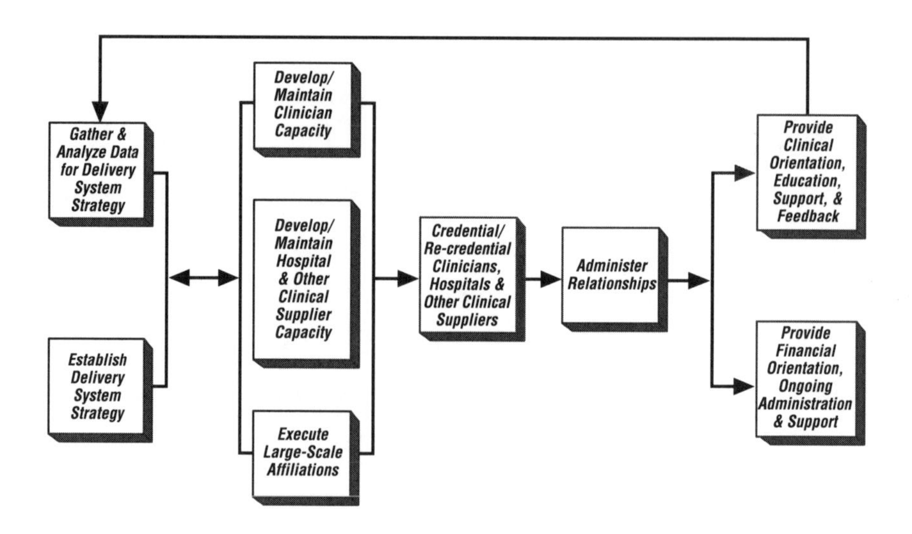

Figure 4-5. Subprocesses Comprising the *Manage Delivery System Capability* Process.

Of particular importance in the design of this process is the need to range far beyond the traditional mind-set of many managed-care organizations that focus on contracting for specialized services. This process—and the new mind-set that accompanies it—will be critical to all types of health care organizations—not only managed-care organizations. It will be the rare health care organization in the future that will stand alone. Most, if not all, will be part of integrated delivery systems that provide services across the continuum of care and link to specialized capabilities such as disease management.

This process defines and sustains clinical capability. It calls for envisioning the scope of services, geographic reach, and clinical capabilities needed by the organization. After defining the core capabilities of their organizations, health care executives will begin to choose the clinical partners they need in order to represent an integrated delivery system.

In addition to letting us aggregate and analyze demographic information to produce a picture of potential demand for services, special needs of populations, and penetration of current services, information technology provides the capability to catalogue the range of care services available in an integrated delivery system and to aggregate information about services provided by each component of the network, including essential credentialing information.

Manage Customer Relations. Similar to triage and advice, the Manage Customer Relations process sits at the center of a diverse array of information—about products, individuals' coverage, available care delivery resources, and billing and payments. As one of the health care organization's support processes (see Figure 4-3), Manage Customer Relations must provide capability to a number of other processes in the organization. In the future customer service specialists will sit before computer screens as complex as the cockpit of a transcontinental jet. They will be able to respond to queries ranging from a change of address to identification of a care delivery resource in a patient's neighborhood. Customer service will be a one-stop source of help to members and patients, purchasers, and providers.

How will that process work? A new member calls for information about the health club benefit included in his group contract and to choose a primary care physician. The customer service specialist refers first to the contract provisions for the member's group and informs him that to join the health club he has selected he will need to pay a small copayment. The service specialist also clicks on a button that sends E-mail describing such additional health and wellness benefits as aerobics classes and suggested routine care appropriate to the member's age.

Simultaneously, the customer service specialist checks to make sure that the member has completed a new-member profile that includes important demographic and lifestyle information. She tells him that the primary care physician he selects will review the background information and contact him with further suggestions for wellness programs or specific preventive care.

Their conversation turns to the second reason for the call: selection of a primary care physician. The member wants a physician near his work, not his home, and, he adds, he wants "a young one who will understand guys like me." First the specialist calls up a list of physicians who fit the member's geographic parameters. Then she sorts them by age. Knowing the age of the member and having asked where he had gone to college, the specialist suggests four primary care physicians who are only slightly older than the member. Again, with the touch of a button, she adds the profiles of the four physicians to the member's E-mail message.

In that scenario, the customer service specialist uses information from many different systems across the organization. Had the member wanted medical information, she could have transferred his call to the specialists providing triage and advice. Aside from that, she had all the information at her fingertips.

Purchasers and providers would enjoy the same comprehensive service. Purchasers might question enrollment status, usage of certain benefits, and payment of invoices. Providers might want information about other available resources.

What makes it possible to provide such a comprehensive service capability is the power of information technology to pull together data from numerous sources. Without that technology, no customer service specialist, however talented, can provide such support. No more slightly out-of-date manuals. No more callbacks and telephone tag. No more manually produced letters. No more frustrated customers.

In the Health Care Infocosm, the needs of members and patients, providers, and purchasers can be met by changes in the customer relations process that are made possible through information technology.

Looking Forward

Whatever strategic choices shape the processes of the future health care organization, one thing is certain: No organization can long survive if it walls itself off from the rest of the industry in the Health Care Infocosm.

In this emerging environment, information technology will enable the forward-looking health care enterprises to open themselves up to the free flow of people, ideas, and information. Old models of health care functions and departments, clogged with inaccessible information, will be discarded. Tomorrow's health care enterprises will look for ways to access the riches of knowledge that the industry is mining, as well as ways to share its own wealth of knowledge with other participants, even those who once stood as formidable competitors. As the participants in the Health Futures Forum in Australia recognized, the use of information technology will be truly transformational, and it will announce the arrival of a new industry structure dominated by integrated and virtual organizations, eager to accept the challenges of managing health care and keeping people healthy (see Figure 4-1).

The great promise of the Health Care Infocosm lies in its ability to integrate the work of the entire industry, keeping consumers well-informed and enabling health care professionals to keep people healthy as well as to deliver the highest quality care when and where it is needed. As internal and interorganizational processes attain unimaginable levels of efficiency, so the work of future health care professionals will change dramatically in the Infocosm. In the next chapter, we will look more closely at some of these changes, alterations that not only will define the jobs that people will do and the attitudes that they will bring to their work, but also will establish the relationships that future health care enterprises will enjoy with their employees.

CHAPTER 5

WORKING IN THE INFOCOSM

Thanks to powerful technology, information now zooms across continents and time zones, linking people together in ways that allow them to share data and to communicate on their own "time." We still sleep and wake by the age-old twenty-four-hour cycle, but we work by clocks that don't know A.M. from P.M., Saturday from Tuesday, holidays from workdays. Just as it does not matter whether we work in Miami or Melbourne, in the front office or at home, the revolution in information technology is transforming the health care industry by reordering the relationships of the health care enterprise. An emerging set of cooperative and collaborative relationships requires all of us to learn to work and manage in entirely new ways.

As health care enterprises transform themselves in response to the new assets information technology creates, that imperative—for collaboration and cooperation—becomes a critical element of required change. The introduction of information technology as a strategic business resource both levels management hierarchies and puts information where people need it for their decision making: on the front line with care providers, product managers, and others who, like them, do the work of the health care organization.

In this chapter, we examine the human and organizational impact of the emerging Health Care Infocosm. Yet the types of changes mentioned in this chapter are the hardest to achieve; harder than getting the strategy right, harder than using technology, harder than redesigning processes and forming alliances. Those who work in the Infocosm will face challenges to very strongly held beliefs about people, organizations, and value. Managing the introduction of change may be the most significant task facing health care executives.

In "virtual" health care organizations that provide care across a continuum, information and ideas flow freely. Virtual enterprises are more nimble, harness more resources, and achieve a greater responsiveness to the marketplace. The key to successful virtual enterprises is the link between these companies and their customers.[1]

The traditional health care model cannot respond adequately to this virtualized world. Indeed, the people who run health care organizations will have to adopt new management models to operate in the networked world. Such a model one of our colleagues has dubbed a "web of capability."[2]

Virtual organizations, which are in response to the "outside-in" demands of customers for broader and more comprehensive health care products and services, will depend more on *infostructure* than on infrastructure. The very survival of virtual health care enterprises will depend on their ability to convey information to the people who create new knowledge and who make decisions that affect the quality of their members' and their organizations' lives. Imagine the potential for improved health care when information flows freely within and across organizations. But to unleash that power, people in the health care organization have to adopt new ways of operating, create new incentives, and develop new assumptions about their work in the transformed enterprise.

People's knowledge and their ability to access diverse stores of information will differentiate competitive organizations in the Health Care Infocosm. That means finding the right people for the right jobs, enabling those people to refine their skills, communicating clearly about performance and rewards, and helping people to develop new capabilities. Transformed organizations, in other words, demand transformed people.

Throughout the health care industry we are witnessing the shift from traditional, managerial hierarchies to collaborative teams that enjoy a greater sense of ownership and responsibility. Alliances and partnerships will extend employees' problem-solving techniques and their access to information beyond organizational boundaries. Information technology will continue to define not only what work people do, but also where they do it and how they are rewarded for their contributions and skills.

Jettisoning Old Assumptions

Restructuring work and its management is one of the hallmarks of the 1990s as information technology is having an increasingly disruptive effect on the hierarchical organization of work.[3] The pace of progress within traditional bureaucratic structures—huffing and puffing through the heyday of growth—cannot catch up to the blistering pace of the technology revolution. Traditional health care structures have isolated specialists from one another, from the organization, from fellow providers, and from the customer. Health care enterprises in the Infocosm will cast off the attitudes that have restricted their progress.

Everything is changing for health care people and organizations, as illustrated in Figure 5-1. Empowerment replaces the need to seek permission. Knowledge moves to the forefront—ahead of skill—as a critical value. The center of gravity—where work gets done—moves from individuals to collaborative teams, and loyalty shifts from one's company to one's profession. Increasingly broad perspectives replace the narrow specializations that have received such fine-tuning over the last few decades.

With these changing dynamics, the critical need for a clear, operational vision (as we discussed in Chapter 3) becomes essential in the formula for success. Such a vision is a tangible manifestation of the leadership required to direct and guide the health care organization into the Infocosm. It also provides the mechanism for people throughout the enterprise to take on ownership of the resulting strategies and tactics to make the vision a reality. Then enablement in the form of incentives and feedback, for example, must be carefully realigned to mirror the directions chosen. With this realignment, the journey to the Health Care Infocosm can be navigated.

YESTERDAY	TOMORROW
Permission Seeking	Empowered
Skill Based	Knowledge Based
Individual Centered	Team Centered
Company Loyal	Career Loyal
Functionally Trained	Continuously Learning
Specialized Expertise	Cross-Functional Perspective

Figure 5-1. The Changing Dynamics of People in Organizations.

The dawn of the Health Care Infocosm has cast the traditional industrial model of controls and commands into darkness: No longer will supervisors issue orders to supposedly uncritical workers, check their output, send memos up and down the hierarchy, and create a cumbersome bureaucratic machine. In the light of this new age, business will no longer grind on as if time were plentiful, as if competitors would never equip their people with information technology or empower them with confidence for faster reaction. As we saw in the case of El Dorado Health Care in Chapter 2, the successful enterprise will recognize the importance of engaging all its people in the transformation process itself: Both the people and the organization enjoy the benefits of the new model for working in the Infocosm.

Throughout the industrial world—perhaps most evident in Europe and Japan—people have long assumed that a "contract" existed between employer and employees: Good performance promised, if not guaranteed, a lifetime job. Although many challenged it during the 1990s, that unwritten contract has helped nourish employees' attitudes that they work "for" an employer to whom they owe loyalty. Today both employers and employees must recognize that this old contract is no longer valid. One human resource specialist warns that "nowadays employees who hold 'employed' attitudes—the belief that they work *for* an employer and that by doing their work, their jobs are secure—are hopelessly out of touch."[4]

This counsel applies equally to health care organizations that assume that most of their employees are theirs to retain or dismiss as it suits them. Health care professionals are less dependent these days. Accordingly, executives and administrators must view their personnel as a different kind of business asset. Caregivers can exercise their free will to choose from a range of options, one of which is the possibility of working elsewhere. As information technology makes an increasingly greater impact on the management of health, and as people must no longer do their jobs in specific locations, health care enterprises and professionals will forge new relationships with each other.

Another old assumption—that "the boss" will assure the competitiveness of the enterprise—will also shatter as we move into the Health Care Infocosm. Mid-level managers have long assumed that top brass—who enjoy a broad view of the industry and its players—will take care of competitive strategy. Information technology, however, undercuts that assumption by making relevant data about competitiveness and profitability available to *everyone* in the organization who needs it.

In a recent study entitled "The Future of European Health Care," Andersen Consulting reported that competition is not the only driving force. Cooperation is another crucial factor.[5] Many in the health care industry are accustomed to thinking of their organizations engaged in pitched battles with the opposition. The current rush to create alliances, joint ventures, and new partnerships will mean new forms of competition. Former competitors will be part of a large "health system" network that includes multiple hospitals, clinics, physician groups, home health care companies, and related businesses.[6] Partners in alliances and joint ventures like this will find themselves sharing information and cooperating rather than competing.

A New Model for Working in the Infocosm

Having divested the old assumption that all decisions must originate at top organizational levels, care providers responsible for keeping people well and for assessing, diagnosing, and treating illness will need immediate access to all relevant data. The same is true for those with responsibility for the health of an enterprise.

With ready access to both new sources of data and new information itself, people working in the Infocosm find that they must be smarter.[7] Blessed with unprecedented power to acquire and disseminate information, they must reconsider, for example, the concept of "owning" information. Physicians—perhaps acquainted only by a videoconferencing interchange—need to learn how best to share information. With access to vast quantities of digitized medical information, the health care givers of the new era will hone skills they'll need to assemble the puzzle parts in effective ways.

Creativity and Collaboration. Health care professionals have long worked in a collaborative environment. Consider a surgeon and an anesthesiologist: Neither can conduct the operation independently; they must work together to achieve a common goal. Today, as we survey the health care landscape, we see that health care organizations are collaborating with their personnel in new ways. They encourage creative problem solving, initiative, and self-direction that come with independent—and ultimately interdependent—thinking. The information technologies of the Infocosm will reward creative thinking and quick reaction with unprecedented power to acquire and disseminate information. But no health care organization can reserve those rewards—speed, agility, reach, and insight—for itself, alone.

In *The Age of Unreason,* Charles Handy offers a new definition of "success" for the era of information technology:

> The new formula for success, and for effectiveness, is $I^3 = AV$, where *I* stands for *Intelligence, Information,* and *Ideas,* and *AV* means *added value*....In a competitive information society, brains on their own are not enough, they need good information to work with and ideas to build on if they are going to make value out of knowledge.[8]

Success will also be judged by the ability of a health care enterprise to increase its capacity to change, using the information and insight made available through information technology. Health care executives must learn to recognize or create catalysts to trigger change based on such things as clear performance targets and measures. Furthermore, success will be judged by an ongoing commitment to make use of the new assets of the Health Care Infocosm that will create a cycle of continuous collection, interpretation, and response.

Instead of modeling themselves on the organizational philosophies of the armed forces, industrial-age corporations, or government bureaucracies, tomorrow's health care organizations emulate institutions where "knowledge has always been key and brains more important than brawn."[9] And this must certainly be the case of future health care organizations where the art and science of medicine converge. Here is where the true value of information to calibrate decisions comes to light. This capability is dramatically different from the automation of processes that provided efficiencies but not insight.

As each health care organization involves its professionals in the creative solution of its problems, the old bureaucratic structures either crumble in the process or stop progress altogether. In many of the world's high-tech, leading-edge organizations, the word "manager" has already disappeared: Self-directed thinkers and doers have adopted essential roles in the organization's success. Management as an *activity* still has a place in modern business, but as a *position,* it is less meaningful. Indeed, in some organizations, the word *manager* is giving way to the term *coach,* suggesting an individual with the skill to direct, inspire, and assist employees. Employees at El Dorado Health Care learned to see themselves as "partners" in the enterprise's vision. Not only did partnering serve as the basis of bedside care, but it contributed to the feeling that the care providers worked more for the patients than for the organization.

The new economic model for the health care industry assumes that every person must fulfill managerial roles, but no person can act *solely* as a manager. Of course, our assumption postulates intelligent behavior supported by easily accessible information throughout the organization. In another instance of the Infocosm demolishing the barrier of place, interaction—a by-product of the "virtual" organization—occurs in cyberspace rather than in buildings of bricks and mortar. The new model promotes mutually beneficial collaboration.

Successful collaboration most often begins with intractable problems and the people committed to solving them. Health care professionals and their organizations are developing compelling visions that focus on improving service levels to the citizen-patient, elevating the quality of care, and reducing costs along the way. Although they are perhaps unattainable by today's standards, in the future those measures of excellence will find ready solutions through improved information technology. Such visions will result, for example, in physicians' working in settings that make professional, personal, and economic sense rather than having geographic proximity as the determining factor. Far more supple and fluid than the old, industrial-age model, the new organizational paradigm—as we pointed out in Chapter 4—weaves alliances and partnerships among groups with complementary capabilities. The availability of information—across continents and time zones—pushes health care professionals to consider their organizations in terms of cooperation.

Collaboration requires people to share responsibility within groups rather than take responsibility solely as individuals. Open communication, mutual respect, and individual skill are also necessary. But most important, collaborative success—whether in delivering a wellness program or designing new products—requires a coherence with vision and goals. Health care organizations and their people need to *learn* to work collaboratively.

True collaboration embodies such concepts as shared goals, continuous communications, and clear lines of responsibility with no restrictive boundaries. Michael Schrage asserts that "collaboration is a far richer process than teamwork or traditional communication: It is 'the creation of value.'"[10] Information technology is enabling continuous communication in innovative ways. Through the Internet, organizations use E-mail to set up appointments and keep in touch with members who go out of town for extended periods of time.[11] Electronic medical records, bedside monitoring devices, and electronic call centers are making all caregivers equally responsible for the continued health of the populations they serve.

In addition to adopting new communication methods and technologies, health care professionals need to become comfortable in a collaborative work environment, and they may find themselves in unfamiliar territory. Decisions need not be consensual: Collaborators frequently bicker and argue. Rather than friendship or affection, successful collaborations require mutual respect and tolerance. Conflict is appreciated as a form of passionate exploration rather than a behavior to be eliminated.[12] Health care organizations must create shared space for their people, and they need to recognize that sharing both informal and formal activities—playing bridge or climbing mountains together, as well as working shoulder to shoulder in office or laboratories—contributes to successful collaboration.

Personal Careers, Not Corporate Careers. In addition to fashioning its particular vision, revamping its processes, and fostering collaboration, the successful health care organization will redefine itself in relation to its professionals. It will learn to capitalize on the skills of its people and to provide incentives for their productive, collaborative work. To support decision-making, the health care organization in the Infocosm will provide its employees with sophisticated, quick-response information systems like the call center capabilities described for triage and advice that we described in Chapter 4.

This restructuring calls for a new professional contract—one that stipulates that health care organizations will provide their employees with opportunities for self-improvements that will advance their careers.

For their part, caregivers will endorse organizational goals and work to advance them. Organizations are restructuring with an eye to putting people where they belong—namely, at the center of service delivery. This is the essence of empowerment. While they draw important capabilities from individuals, groups, and other enterprises with whom they are affiliated but do not control, health care organizations will need to invent new ways to unite around their compelling visions for the future. Care providers and other professionals will no longer depend on the organization to ensure their jobs; rather, they must be responsible for discovering how best to develop and use their particular skills to accomplish new tasks and solve problems. What will matter—in the profound redefinition of labor relations—is lifelong *employability* rather than lifetime *employment.*

Valuing the Individual's Knowledge

The introduction of integrated processes to manufacturing industries in the 1980s brought about a transformation that is a source of both inspiration and instruction for the health care industry as it confronts tremendous challenges. Combining advancements in manufacturing technology with total quality management and just-in-time inventory control—elements that make the most of innovations in information technology—

manufacturers effectively changed both the assumptions and the practices that characterized the way things used to be done.[13]

In place of the traditional power pyramid, we envision a structure based on the sharing of power, an increased sense of responsibility and accountability throughout the health care organization, and a system of incentives and rewards that recognizes the particular contributions—added value—that each provider brings to the enterprise. Such an environment calls for honest self-assessment, frank exchange of information, commitment to agreements, proficient dispute resolution, and consensus-building skills.

As health care organizations transform themselves, they must devise new ways to value knowledge. Edward E. Lawler, who directs the Center for Effective Organizations at the University of Southern California School of Business Administration, says that "people will no longer be paid for what the job is worth, but, in essence, for the value that they can bring to the corporation and what they are capable of doing."[14]

Following this line of reasoning, a number of health care organizations are experimenting with variable compensation programs. They start by offering incentives first to top executives, but their plan is later to extend the program to all employees. Like many organizations in all industries, Presbyterian Hospital in Charlotte, North Carolina, has recently initiated a program that rewards executives who attain financial goals. Samaritan Health System in Arizona has redesigned its compensation program to recognize the added value that its 11,000 employees bring to that organization.[15] Skill-based compensation can "reinforce employees' efforts to understand an entire production system, thereby enhancing their ability to diagnose and solve nonroutine problems."[16] Such widespread innovations to compensation programs call for carefully aligning individual contributions with organizational objectives. Organizations that want to incorporate the valuable ideas that new knowledge generates must define their compensation programs much more broadly than they did in the past.

To provide the best service, health care organizations will incorporate advanced technology and train their people to work with that technology. If the virtual organization is the new model for organizational structure, information technology is its DNA. Like the study of DNA, information technology poses an intimidating challenge, but health care personnel must do all they can to prevent feelings of trepidation and fear. As information technology continues its breakneck progress, its impact will be less imposing, more routine, simpler to use, and less distinguishable from business processes themselves. For most nurses, physicians, laboratory technicians, and health care administrators, knowing how to work with information technology is more important than knowing how the technology works.

No software program or computer network is worth a dime if it doesn't help caregivers combine their skills and knowledge to achieve their goals. In Chapter 1, we saw how technology is both driving and enabling changes throughout the health care industry. Telemedicine will enable caregivers anywhere to consult with specialists wherever they are and at any time. Today's widespread familiarity with the Internet will encourage people to increase their demands for knowledgeable care providers. Such changes vastly enhance caregivers' facility for genuine collaboration and demand that they be collaborative.

Some organizations are making individuals an integral part of the health care team by encouraging them to become involved in their own care. By opening the lines of communication and providing access to information and education, people have become true collaborators in their own health care.

The most successful *networks* of organizations—alliances and partnerships—will coordinate so well that citizen-patients will no longer detect boundaries between them. A 1994 review of recent developments in architecture concluded that continuing success will depend largely on the "ability to integrate different kinds of knowledge, to pool resources, to synthesize."[17] Ditto for health care. In fact, the need to pool resources and synthesize is more crucial in health care than in many other industries, just because its elements—for example, physicians' knowledge and experience, medical records, laboratory results, administrative procedures—are so extensive, disparate, and diverse, yet need to be so broadly shared.

While knowledge has always been at the center of prevention and health care, the uses, directions, and goals of knowledge are now in a period of intense flux. The restructuring of the industry generates a new economic model whose "currency" is improved health outcomes in the broadest sense: better overall health for larger populations. Admittedly, such concepts aren't easily defined: Precisely what is "overall health," and how can we measure it? Now more than ever, payors, employers, government agencies, and insurers scrutinize the health care system in terms of costs and benefits, and they are all clamoring for consistent definitions and standards. Nevertheless, uncertainty about appropriate measures in no way invalidates quality—

judged by the caliber of individual care—as the key to the survival of health care organizations. If for no other reason than greater efficiency, what shapes the new economic model is concern for the health of the populations that organizations serve. That concern is, therefore, the basis for valuing knowledge.

In health care, competitive and economic pressures and the demand for high quality services increase the need to link physician incentives and those of other care providers to broad organizational goals. The health care industry accounts for 14 percent of the gross domestic product of the United States—significantly more than in other advanced countries. And fully 65 percent of health care's expenditures are labor-related: Personnel costs in the health care industry already exceed those of most other service industries.[18] To control such costs, future employees in health care will use technology in ever more imaginative ways. The tools they need are already available. In the next decade, we will increase computing power and communication speed by astronomical amounts.[19] Would a person choose a physician equipped with this power over another who lacked access to the same "global intelligent network"?

The "predominant information appliance will no longer be a telephone, television, or a computer, but rather a telecognitor—a terminal connected to the global intelligent network—whose primary function is enhancing human cognition."[20] Many of us feel overwhelmed by the gigabytes of information already at our disposal. In a world where information processing takes no time at all at the same time that the amount of information continues to grow exponentially, we will all have to develop the means to access, assimilate, and use the information available to us. Physicians, in particular, face a Herculean task as they try to keep abreast of developments in their fields:

> Ten years ago, medical education was based on the concept that the most knowledgeable clinician was the one who could contain the greatest amount of information in his or her head. Today, as Donald Lindberg, director of the National Library of Medicine, stated, an extremely diligent physician who reads two medical journal articles a night would be 800 years behind in his reading by the end of the year, given the exponential expansion of medical knowledge today.[21]

Coupled with the increase in information available today, economic trends and political developments continue to have an impact on health care professionals. Even physicians, whom the public perceives as a group of professionals blessed with high salaries, are experiencing an unprecedented squeeze on their earnings. The health care industry is, of course, subject to the same pressures other industries experience, and those pressures are certain to increase, particularly as the aging populations in the United States and other countries require more and more health care.

Whatever new compensation schemes are adapted, physicians and others in the industry will need to adjust their training and their assumptions about collaboration to work effectively in the Infocosm. McDonald,[22] for example, argues that information technology will transform the delivery of health care as "intelligent agents" (i.e., software-based assistants) direct health care professionals through mazes of information in global databases. Collaborative decision-making will also encourage them to work in teams, more frequently and profitably than they do now.

New Incentives for New Knowledge and Skills

We must learn to define knowledge in terms of its contribution to outcomes. What is the value of knowledge in incentive and compensation schemes? Would chemists produce more effective pharmaceutical compounds if they earned a percentage of new drug sales? Would nurses in an intensive care unit attend special training classes on the latest information technologies if their newly developed skills were rewarded with increases in their salaries? To tackle such questions, we need reward and performance systems that will help us manage the work people do and systems that can link knowledge with desired outcomes, account for that connection, and provide incentives that stimulate the creation of knowledge.

How will incentive structures actually affect the behavior and activities of health care organizations and their individual members? How can we measure and track individuals' contributions? How can we assure accountability in an alliance-based workforce? Can we permit an individual—even as a member of a team—to file a patent for an invention? If that last question seems outside the realm of our attention, consider the recent case of Dr. Samuel Pallin, an Arizona ophthalmic surgeon. Claiming he invented the procedure for stitchless cataract surgery, Dr. Pallin filed a patent infringement suit against another ophthalmic surgeon in Vermont, prompting intense controversy throughout the medical community.[23]

While there is no precedent for one doctor paying royalties to another, the issue of ownership of intellectual capital in health care is quite complex. Witness the number of such cases at major research universities that seek to control rights and profits from discoveries made by their researchers and teachers. Advances in information technology and our ability to share those discoveries complicate the extent to which medical personnel make use of new knowledge or restrict access to new techniques.

No sooner do we settle one question to our satisfaction than it either changes form before our very eyes or an entirely new question suddenly arises in its place. Still, one thing remains certain in the midst of all the turbulence: The fundamental social contract between health care organizations and professionals will evolve as the Infocosm emerges. In the Infocosm, victory will belong to the most creative application of the best-informed intelligence. All other relationships flow from that central principle. The successful application of intelligence will be determined not by organizational function but by the *functioning* of that organization as a whole. The service a surgical team provides, for example, will be the basis of the team's evaluation: How well the members of the team work together will determine the effectiveness of their service. It will no longer be adequate to talk in general terms about pay-for-performance because "performance" alone will no longer be the key organizational outcome. Measurements of health care will be expressed in terms of value to the citizen-patient.

One practical way for employers to signal their high esteem for both intelligence and their employees is to provide those employees with opportunities for skill enhancement and knowledge enrichment. Such mutually advantageous programs benefit employers because their employees sharpen their skills. Employees, for their part, find gratification both in their salutary effects on people's lives and their awareness that their skills and knowledge are more highly valued in the marketplace. Most important, employer and employee have strengthened their *relationship* by building it on recognition and appreciation rather than control and command, on the real benefits of collaborative work rather than on the safe fantasy of paternalism. The conditions of their relationship free both sides to operate for the better health of patients and members of health maintenance organizations.

As health care professionals find their way in this new world of information technology, experiencing its impact on their delivery of health care, their members will also see and recognize the changes. Although most citizen-patients will remain unaware of innovative incentive plans that the health care organizations institute for their caregivers, the members will be the indirect beneficiaries of those plans. But what exactly will the people of tomorrow experience in the Health Care Infocosm?

In the preceding chapters we have examined information technology primarily in terms of its impact on health care organizations and caregivers. We have seen how it will help to break down the barriers of time, place, and form, and thus extend the reach of health care professionals, who will find and be rewarded for creating new knowledge and providing better service. In the next chapter, we'll shift perspectives and look at the Health Care

Infocosm from the point of view of the individual—in this case a man who has assiduously avoided dealing with the medical establishment. For years our fictionalized Australian athlete, Joe McAlister, saw no reason to worry about his health, and he successfully kept himself out of doctors' offices. Ready or not, at age 50, he now finds himself confronting his own mortality and about to embark on his own voyage to better health care in the Infocosm.

CHAPTER 6

LIVING IN THE INFOCOSM

Joe "Kanga" McAlister,* a star football player for Australia in the 1970s, got his nickname from his deceptive, kangaroo-hopping foot movements just before he scored a goal. Good-natured and extroverted, always responding with an easy smile and a "G'day Mate," Joe was admired for having as much fun off the field as on it. Although he is no longer so quick on his feet, his reputation for deception still fits him, for he is a man of many denials.

Thirty-five years ago, Joe started to smoke, like many of his teammates. Although Joe would never admit to being hooked on the stuff, his addiction to nicotine is now older than his eldest son. He'd much rather talk about his new granddaughter than about his own medical condition. The truth is that he has visited his family doctor, a general practitioner named James Sayle, only twice in the past 11 years.

Now, in 2001, his denial has finally caught up with him. No longer able to explain away his weight gain and low stamina, Joe resolved a few months ago to get back into shape. He took up jogging—intermittently—in the fond hope that he could still do the old Kanga steps. Starting out on his route in suburban Melbourne one evening, he felt an unpleasant ache and tightness in

---

* Joe McAlister and all of the other characters and places in this chapter are fictitious and any resemblance to actual people or places is coincidental and unintentional.

his chest—a more intense version of the discomfort he had felt after walking up two flights of stairs to his office. "Just a little indigestion," he told his wife Michelle.

She, however, recognized the danger signs and insisted that he call his general practitioner for an appointment. The thought of confronting Dr. Sayle made Joe's chest tighten even more: He recalled a visit back in 1992 when he had waited an hour and a half in the reception area and provided the same information he was sure he had provided before. "About as pleasant as having a tooth pulled," he had told Michelle when they went out that night for dinner. Not having seen the inside of a doctor's office for many years, Joe is unaware of the vast changes that have occurred in the world of health care and medicine. He is about to encounter not only his general practitioner, but also the latest technology in the Health Care Infocosm. In short, Joe McAlister is in for a big surprise.

The New Age of Health Care

The first change was evident when Joe called Dr. Sayle's clinic to report his symptoms and to make an appointment: There was hardly any wait. Using a computerized scheduling system, the clinic staff recorded Joe's complaint, matched his preferences with Dr. Sayle's availability, and advised him to avoid strenuous exercise in the meantime. His appointment with Dr. Sayle was set for 9:30 the next morning.

What wasn't evident to him would have astounded Joe even more. While he was describing his symptoms to the clinic staff, the computer was pulling up all the reports of Joe's previous visits—not just those to Dr. Sayle's office, but also from every other health care facility he'd visited in Australia and the one in New Zealand when he'd suffered a dislocated shoulder in a football game. A warning system, enabled by new information technologies, signaled an alert to the clinic staff by highlighting the results of Joe's last visit, back in 1992. During that examination Dr. Sayle had found that Joe's blood pressure was mildly elevated, and he had ordered a routine investigation of blood lipid levels. Despite several reminders from the clinic, Joe kept putting off the blood test and ultimately failed to get it.

Even with his spotty medical record, Joe McAlister does not escape the discerning eye of information technology. Evaluating Joe's known medical condition, his reported current symptoms, and his lifestyle, an on-line research service constructs a composite picture of Joe and matches him with thousands of other men in his age group who have similar profiles. It searches its reference bank, suggests several diagnoses and treatments, and awaits more data from Joe's visit the next day. The computer relays all the information to Dr. Sayle's electronic mailbox instantaneously, lights an alert icon on his computer screen, and records a voice message on his telephone answering system. The message will be delivered to Dr. Sayle when he is driving to work in the morning.

The General Practitioner as Facilitator

Sitting in Dr. Sayle's waiting room the following morning, Joe notices that things have changed since his last visit. The shelves on the wall behind the receptionist's desk, formerly clogged with files of yellowing medical records, have been replaced by two large watercolors. Where there used to be two busy receptionists, now only one sits at the desk, apparently talking into his computer. There is no paper to speak of anywhere in sight. Expecting a crowded, noisy reception area, Joe is surprised to find the waiting room quiet. Two people sitting in a nook on the other side of the room are engrossed in front of a computer labeled, "To Your Health—the CD-ROM Library."

Dr. Sayle appears, shakes Joe's hand warmly, and invites him into the consulting room. Joe is embarrassed when his friend reminds him that he never had taken the blood test ordered in 1992. Although Dr. Sayle had not installed a full electronic record system when Joe last visited, he had since computerized his notes, lab results, medications—all the information relevant to Joe's physiology. Now Joe's record appears on the computer screen like an accusing finger, reprimanding him for too much machismo and too little concern for his health. The key results of his previous exams and Dr. Sayle's unambiguous instructions for tests are highlighted in bold.

Reluctant at first, Joe describes his indigestion, chest pains, and the slight swelling of his ankles at night. Dr. Sayle occasionally responds with what seem typical queries: "Hmm...yes, go on....Have you experienced any other discomfort?" Joe notices that instead of scribbling notes on a piece of paper, his friend dictates into a microphone on the side of the computer screen, and at other times touches the screen with his pen. More questions follow: "How much exercise are you now getting in addition to the infrequent jogging? How many flights of stairs can you climb before you feel the tightness in your chest? Do any specific foods worsen your indigestion?"

After weighing Joe and taking his blood pressure, Dr. Sayle asks Joe more questions, some of which are prompted by the computer in the examination room. They talk about the medical history of Joe's family, focusing especially on how and at what age his father died. Joe responds that his father, who was also his hunting and fishing mentor, died in his sleep during a trip to Outback Australia. He was only 47 years old, three years younger than Joe is now. The diagnosis was heart disease.

Dr. Sayle tells Joe that his symptoms of discomfort and the record of his father's heart disease suggest a problem with his own heart. Accordingly, he recommends further investigation. The doctor types in a request for an EKG and some baseline blood tests and prescribes nitroglycerin tablets for Joe to put under his tongue should he ever again feel chest pains. He also instructs Joe on the multiple ways that cigarette smoking is contributing to his problem, and hands him a brochure listing three "Quit Smoking" programs. Joe promises to follow up immediately.

On the spot, the computer generates a reminder for Joe to pick up the prescription, a detailed description of the indications of heart disease, and recommendations for diet and exercise. Joe is happy to have the printed information because most of Dr. Sayle's explanation escapes him. Actually, he isn't listening. Until now, he hadn't thought much about his mortality, even though he must have suspected that the chest pains were probably indications of something serious. The prospect of heart disease rattles him.

Before he leaves Dr. Sayle's office, Joe signs a national health insurance voucher. The receptionist explains that if he had brought his health card with him, Joe wouldn't have needed to sign that form, but his forgetting is hardly a serious problem: The receptionist simply consults the region's database, locates the universal patient identification number (UPIN) assigned to Joe by the Federal Health Department, and prints a form confirming the consultation and the charges. Every evening, the receptionist forwards the daily compilation of Dr. Sayle's transactions in a batch to the regional health authority offices for processing. Joe's record will, of course, be included in that day's file. Payment for Dr. Sayle's services are electronically transferred to his bank account within twenty-four hours.

When Joe arrives at the pharmacy, his prescription is ready. Dr. Sayle's computer had sent the order directly to the pharmacy so no paper was involved—except the reminder Joe has crumpled in his hand. When Joe gets his nitroglycerin, he also notices that the pharmacy computer has a complete record of every prescription Joe has received over the past two decades. The pharmacist explains that she depends on the information system to highlight potential drug interactions at the time of prescription and alert the physician to side effects or the potential for allergic reaction. It also prints out the relevant consumer information about the drugs each patient receives, including correct medication instructions, potential side effects, and an explanation of what the drugs actually do. Finding all this difficult to take in, Joe asks her how she ever got along without this system. Amused by his naiveté and shocked by the prospect of doing without the convenience and accuracy of the system, she responds, with utter honesty, that truly she had no idea.

His suspicions raised about Joe's health, Dr. Sayle knows that his computer screen will flash a message in red and sound an alarm if the test results indicate the need for immediate response. To Sayle's relief, no such message appears. Checking routine results on his terminal at the end of the day, he sees that Joe's EKG shows nonspecific T-wave charges, but no other abnormalities. Joe's elevated serum cholesterol level does seem cause for concern, but all other results look quite normal. While looking at the results, with the simple touch of the screen Dr. Sayle creates a new order to have Joe get another test to rule out heart muscle damage and asks his receptionist to send Joe an E-mail message to come in first thing in the morning for the follow-up test.

Before he has a chance to read his E-mail in the morning, Joe arrives at the clinic at 8:30, complaining that he felt severe pains in his chest just before he sat down to breakfast. The nitro tablets—he had to take two of them—relieved the discomfort, but now he has a splitting headache. When Dr. Sayle persists in his interview, Joe reveals that he was worried about his lack of stamina, and had decided to deal with the problem himself. True to machismo form, Joe had decided a more serious jog would whip him back into shape. When he developed severe chest pains before he reached the end of the first block, he had stumbled home, the pain spreading into both arms.

Dr. Sayle suppresses a groan. Former athletes can be so difficult! Reminding Joe that he warned him yesterday not to attempt anything strenuous, he orders an EKG and runs the test for muscle damage. Both look normal, and Dr. Sayle sends Joe home with more vigorous orders to cut out the jogging and, for that matter, all other exercise.

Reading the anxiety in Joe's eyes, Dr. Sayle explains that the pain is a typical symptom of angina pectoris, a condition that occurs when arteriosclerosis or other types of coronary heart disease constrict the blood vessels beneath the sternum. To illustrate the problem, the doctor turns to his computer, calls up a lifelike color picture of the heart, and rotates the screen for Joe to see. Pointing to diagrams of arteriosclerotic blood vessels, he gives Joe a graphic description of coronary disease as it progresses through the arteries. The doctor next taps into the appointment schedule of Dr. David Dougan, the Melbourne cardiologist to whom he prefers to refer his patients. The computer reports that Dougan has no open slots in his calendar until the following week. Unwilling to wait that long, Dr. Sayle sends an urgent message to explain the situation, and Dr. Dougan quickly responds, scheduling a consultation the following day. Dr. Sayle is comfortable with a one-day wait, because his computer confirms that according to a recently approved best practice available, a negative EKG and a negative blood test indicate that imminent serious complications are highly unlikely.

Dr. Sayle asks Joe if he has any objection to sending his medical record to Dr. Dougan. Far from objecting, Joe wants to thank his old friend for moving things so quickly, but he finds he can only nod his head and let fate take its course. Dr. Sayle, who relies on information technology rather than on fate, selects details from Joe's files, and forwards them to Dr. Dougan's office through the health care network. Before he makes the instantaneous transfer, however, Dr. Sayle first uses his computer's voice-understanding capability to dictate a note to Dr. Dougan and attaches it to Joe's file. Simultaneously, that information goes to the medical history database, which stores summary information on individuals. The demographic and medical information the database stores is useful for regional and national planning, especially preventive planning. Once the information has been stripped of any ties to Joe, appropriate medical researchers and practitioners, as well as public health officials, can access the data for research and analysis.

When Specialists Already Know Your Name

Dr. Dougan starts his early mornings by reading medical histories from referring physicians. Instead of a barely decipherable "Dear Dr. Dougan, Thanks for seeing Mrs. Jones, who has been having chest pains," he reviews clear, complete summaries from doctors who easily use computer programs that prompt them to answer questions Dr. Dougan will have about their patients. He knows that because he starts out with the essential facts about each new patient's medical condition, he'll be much more effective and he'll also avoid wasting valuable time asking repetitive questions. He's confident that he has a comprehensive view of each person's history, including important details some people might forget to mention.

Opening Joe's file, Dr. Dougan reviews the medical history and notes his athletic career. Dr. Dougan listens to Dr. Sayle's comments on Joe's history, symptoms, and signs while he studies the written record of Joe's family history, the EKG, and blood test results. All the information confirms Dr. Sayle's suspicion that Joe has angina pectoris. A stress EKG, however, would convince Dr. Dougan.

Following his initial discussion with Joe, Dr. Dougan asks Joe to begin the stress EKG. But halfway through the exercise test, Joe experiences severe chest pains and has to stop, a clear indicator of significant coronary artery disease. Dr. Dougan is concerned that Joe's angina is unstable, an indication for urgent coronary angiography.

Dr. Dougan looks on-line for the available times at St. Anthony's, an ambulatory- and acute-care center, where such procedures are performed. Finding that the reserved emergency slot is still open later that day, Dr. Dougan enters Joe's identification number and, automatically, he forwards Joe's preadmission information to St. Anthony's and to Dr. Sayle who wants to be kept abreast of Joe's situation. Like many of his colleagues, Dr. Dougan at first had grave doubts about the computerized network, but after many cases like Joe's, he is convinced that things go more smoothly under the new system and that the breakdowns in communications don't happen the way they used to with consultations done by phone and medical records moved by messenger.

Joe's angiogram reveals severe vessel disease. Dr. Dougan meets with Joe and his wife in a consulting room. He explains the test results as well as what they can expect to happen now. Joe's diagnosis is angina pectoris which, the doctor explains, is reduced oxygen supply to the heart muscle, usually caused by a narrowing or obstruction of the coronary artery. That obstruction accounts for the viselike pain Joe felt below the breastbone when he sprinted. Sometimes, Dr. Dougan explains, arteriosclerosis responds to drug therapies, but in Joe's case, he recommends a procedure called an angioplasty. Dr. Dougan describes the procedure in general terms and shows the McAlisters how to download additional information from the World Wide Web: details about the diagnosis and procedure and a discussion forum with other families who have faced the same procedure.

When Michelle expresses anxiety about her husband's chances for survival, Dr. Dougan refers to the nearby medical university hospital's database and matches Joe's profile with a large group of men with similar conditions. To illustrate the conclusions he is sharing with Michelle, the doctor reviews the outcomes those men experienced and explains the relatively low risk associated with this procedure. He also shows the independent review of the center's overall excellent outcomes compared with other centers, both regionally and nationally.

He has his computer play a full-motion video of the procedure and related graphic presentations. From time to time he stops the video so that he and the McAlisters can discuss what will happen in Joe's case. Dr. Dougan reviews the care plan that Dr. Li, the cardiac surgeon who will do the procedure, favors.

Although he is still worried, Joe feels better for understanding what to expect. For his part, Dr. Dougan wonders how physicians used to cope with the enormous volume of information before computers had been integrated into the health care industry.

As he drives to an appointment at another facility, Dr. Dougan uses his car's voice-activated computer to forward Joe's records, along with an explanatory note, to Dr. Li.

Seamless Care

As Dr. Sayle monitors Joe's progress through updates to his medical record, he reflects that while care delivery is more specialized, medical advances and treatments have astounded everyone in the health care industry. But there is also a tendency to fragment care. When he needs to refer people to several care providers who work within different organizations, he worries that they might feel overwhelmed and lost in the system. Not long ago, the family doctor was the integrator and interpreter of an individual's lifetime medical care, and Dr. Sayle himself has always tried to deliver care in the framework of stable personal relationships with his clients. There was a difficult period during which the general practitioner was often out of the loop: People often bypassed general practitioners like Dr. Sayle, going directly to a cardiologist or dermatologist for treatment. In the Health Care Infocosm, Dr. Sayle now has regained the role of captain, and he is central to the care team wherever and however his patients are being treated.

Dr. Li and the others at St. Anthony's also enjoy the benefits of that collaborative synergy. When he reviews the day's calendar, Dr. Li sees that an urgent case has been scheduled for later in the day. He reviews Dr. Dougan's message and Joe's medical history, along with all the test results. He makes sure to set aside time to meet with Joe as soon as he arrives at St. Anthony's.

Dr. Carson, the registrar for the cardiac unit, has both a medical and a business degree. He also reviews the request for Joe's admission. A specialist in information technology, Dr. Carson manages a complex department with connections throughout the health care network. His computer's large interactive display lets him review all admission requests, and he checks each member's medical history, test results, and physicians' orders for treatment.

The cardiac unit has devised a rule-based algorithm to help the doctors rank the cases in order of their priority: That's the mechanism that set aside the "urgent" slot for Joe's procedure.

To understand the demand for the cardiac unit's resources, Dr. Carson takes note of the customized care plan that Dr. Li has ordered for Joe. The care plan allows Dr. Carson to anticipate other services that Joe will need in the course of his recuperation at St. Anthony's. Furthermore, the care plan indicates follow-up and rehabilitation care after Joe's discharge.

The Care Unit manager reviews the pending procedures list for the day. Noting Joe's condition, she assigns him a patient/caregiver dependency rating and scans schedules to select the care team that will follow his progress throughout his stay at St. Anthony's. That assignment also employs rule-based algorithms to help select the most appropriate care team available for each shift. The flexible system flags potential scheduling conflicts and proposes the optimal combination of caregivers.

To further complicate his anxiety, Joe has been feeling guilt pangs: He's realized that his carelessness and bad habits have hurt him and his family. Conscious of his own mortality, Joe has lost his cocky, football-star attitude, and he's depressed. Michelle, who did read the materials that Dr. Sayle had provided, was prepared for Joe's emotional reaction. Aware of his anguish, the care team members also know they can't focus on the clinical details alone.

Joe's angioplasty and the removal of blockages from his arteries proceed without complications thanks, in part, to a laser procedure that is only minimally invasive. That advanced procedure is faster than balloon angioplasty and has a low complication rate. Because Joe did develop a slight fever and cough the day after surgery, Dr. Li extends his stay in St. Anthony's cardiac monitoring center. But with antibiotics, respiratory therapy, and the T.L.C. of the center's staff, Joe goes home three days after his angioplasty.

At Home: Under Doctors' Watchful Eyes

Joe left St. Anthony's armed with a rehabilitation and diet regimen designed to get him back on his feet. Like the admissions procedure, discharge is a carefully planned aspect of the care plan. The goal is to coordinate care both in and out of the acute-care setting. All participants in Joe's case are involved: the pharmacy, a rehabilitation facility near Joe's home, a dietitian who will counsel Joe, Dr. Dougan, and, of course, Dr. Sayle.

When Joe leaves for home, St. Anthony's updates his smartcard with the discharge summary and other pertinent information. The McAlister family can feel confident that, in the case of an emergency, any doctor or paramedic will have immediate access to Joe's current medical information, either through his smartcard or the medical history database.

As if Joe weren't already amazed by the changes he's seen in health care, he finds that the Infocosm has followed him home. In the familiar comfort of his own bedroom, Joe simply turns to his bedside table, where Michelle has placed Joe's laptop computer, and he has instant, on-line contact with the Care Team back at St. Anthony's. The team, at regular intervals, prompts Joe to take his medication, to check his blood pressure by placing his index finger against a touch-sensitive device connected to his computer, and to describe his general physical condition.

As his recovery progresses, Joe frequently goes onto the Internet to converse with other men who are in the midst of similar recoveries. The men compare notes and give each other support, both in progressive activity and in staying off the cigarettes. When he is strong enough to go for prescribed workouts at his local gymnasium, he uses a computer hookup to connect to his Care Team and record his exercise progress. The team is careful to keep Joe from pushing himself beyond the bounds of Dr. Dougan's advice.

The Many Beneficiaries of the Health Care Infocosm

Joe regains strength and his life takes a new direction. He carefully follows the regimen laid down for him on discharge: He takes his medication on time, profits from physiotherapy and rehabilitation services, exercises only as directed, follows the nutrition plan, and is continually in electronic touch with his Care Team. And, of course, he successfully and finally quits smoking.

Although Joe was enormously impressed by and grateful for the smooth integration of his treatment—the lack of jumps and starts between phases of treatment, the relief from repeatedly reciting his medical history, and the remarkably coordinated teamwork—he understands little of how it has all come to be. Almost nothing about the Health Care Infocosm or the activity that supported his treatment is visible to him. His ignorance is understandable: The workings of the Infocosm are transparent. Efficient running of a care center in particular and entire wellness or care process in general is not the concern of the individual.

Throughout his treatment, Joe has marveled that the doctors and the Care Team have made him feel like the person who most mattered in the entire episode: a valued customer whose business specialists court by providing good, quick, reliable service. How times had changed since his last foray into the medical community.

Medical Records, Communications, and Care Plans

There are no scribbled notes shoved into Joe's medical file, there are no messages that doctors or other caregivers can misread or lose, and nobody has missed messages or spent hours returning phone calls to no avail. The comments and remarks of every professional involved in Joe's treatment are accessible in a file that everyone shares. In the past, health care professionals might have considered themselves members of a team by virtue of their common goal: the resolution of medical problems. But everyone worked in a different arena, isolated by the inconvenience and delays of written and telephone communication. In the Health Care Infocosm, the exchange of information is smooth and global. Teams are easily linked through networks of computers—not simply to treat people but also to explore the most recent discoveries compiled from research the world over.

In addition, team members have a sense of themselves as participating in a continuum of care: Physicians, support personnel, and institutions act as partners with shared goals, rather than potential competitors. Psychologically, emotionally, and professionally, they view medicine as the art and science of teamwork.

Beyond that, Joe's case becomes a part of a rich repository of experience related to heart disease in general and angioplasty in particular. The information collected in the care plan is aggregated with that from thousands of similar cases. Dr. Li, Dr. Carson, and the Care Unit manager consider the care plan assigned to Joe a valuable tool for planning and executing treatment. When those procedures were first introduced, some doctors worried that the care plans represented restrictive "cook-book medicine" that oversimplified the complexity of care. In time, however, caregivers came to see care plans as a way to formalize and streamline work they'd been doing all along.

Team members can use computers to call up the care plan for simultaneous viewing, with such documents as medical history, order communications, and test results gathered by the clinical management system. A regional Medical Standards Committee periodically reviews and updates care plans, based on overall outcomes tracked through a longitudinal clinical repository. That review improves caregivers' efficiency and, when unnecessary duplication is eliminated, ultimately cuts costs.

As registrar of the cardiac unit, Dr. Carson has access to all the relevant information; he knows which equipment the surgeon will require, and he can book Dr. Li's equipment requirements in plenty of time. Bar-code readers support the clinical costing system and inventory maintenance. They track and identify all stock items the procedures will use.

Care plans can be directed to activate orders for medication and services, automatically initiating prescheduled orders that remain pending until Dr. Carson or an appropriate duty doctor intervenes. That system has significantly reduced entry, documentation, and turnaround times for doctors' orders. To register orders outside a care plan, the doctor can access order communications through "quick-key" functions or by using a voice command that the computer recognizes. An "intelligent-assist" application provides relevant drug information, displays the cost, and identifies generic substitutions if they are available.

In the past, caregivers spent between 25 and 40 percent of each day preparing documentation and wading through a variety of printed care-delivery references. By the time that Joe McAlister will have his heart problem, on-line care plans will have reduced that time by half, allowing the staff to spend more time with people addressing their concerns. Much of the required documentation is extracted from the results of care plan activities, and team members record detailed notes only in exceptional cases. Most data fields are standardized, providing caregivers on-screen choices from menus. If Dr. Li, for instance, must record notes on an unusual occurrence, he can write on the electronic pad by hand or speak into its recorder. Either way, his comments will be preserved as a permanent part of the electronic medical record.

Collective Experience

In the Health Care Infocosm, Joe recovers, and his experience becomes part of a larger statistical database. His record joins the clinical data in the appropriate sector of the enormous database. The system has been designed to thoroughly safeguard privacy and maintain anonymity. As a teaching facility, St. Anthony's emphasizes research. In its pursuit of better clinical outcomes, St. Anthony's makes its database available to all clinicians and allows for tracking a variety of indicators and relationships. For example, a second-year cardiology resident searches two years of accumulated data to track the effectiveness of an expensive new drug. She is comparing readmission rates of people using that drug to readmission of those on other therapies.

St. Anthony's director of Cardiac Services benefits from aggregated information when he reviews operational reports—especially the daily compilation of activity in all units of the cardiac service. The detailed information helps him forecast demand for service, and his long-range plans coincide nicely with actual need. Any major variations from Care Plans are tagged in red, and he investigates those situations that warrant his attention.

St. Anthony's financial vice president follows daily reports of clinical costs: The system automatically generates income statements for every service. She can monitor and confirm budget and planning assumptions on a day-to-day basis.

From clinical and financial viewpoints, both executives can now look into the future and review the recent past with a good deal more certainty. They can accurately assess overall demands on human resources and facilities, and, as a result, every unit's operations grow increasingly efficient.

Both executives also benefit from the Performance Measurement System, which supports the tracking of key performance indicators identified in the St. Anthony's annual planning process. Its integrated, real-time platform tracks performance in a broad range of areas including clinical quality, customer service, continuity of care, staff skills, and customer satisfaction. Joe McAlister's record is but one in a very large system, but all the Joe McAlisters together have an enormous significance for St. Anthony's as well as the entire Health Care Infocosm.

The Performance Measurement and clinical record systems are critical not only to St. Anthony's and Joe's whole care team. The Performance Measurement System tracks overall quality of care indicators mandated by the Quality Management Program of the regional Health Department, to which it is directly linked. That system supports the "scoreboard" display of St. Anthony's own vital signs and provides a yardstick by which prospective customers and caregivers alike can judge its effectiveness.

The Bottom Line

Our sketch of the Health Care Infocosm isn't simply a daydream: It's what the future holds for smart health care organizations as information technology breaks down the barriers of time, place, and form. Joe's care progressed easily among many members of the health care team who were all focused on the best outcome: Dr. Sayle, the two cardiac specialists, and the teams of caregivers within St. Anthony's and at home. Like well-coordinated players on a winning football team, the health care planners worked together toward one goal—making Joe healthy. Throughout that process, the Health Care Infocosm touched everyone: Joe and his family were well informed and had ongoing information and clinical support. Dr. Sayle, the "family doctor," was a touchstone for Joe and Michelle from start to finish.

Everyone at St. Anthony's focused on care rather than administrative problems and record-keeping. The information about Joe's condition, diagnosis, tests, surgery, and recovery augments an ever-expanding database of medical experience that allows medical practitioners to learn and continually fine-tune their skills.

The bottom line was about bringing the best knowledge and expertise together, independently of where relevant information resided, to identify and manage an episode of health based on Joe's complete needs, achieving the best outcome—every time.

Chapter 7

EPILOGUE: THE CHALLENGE OF THE INFOCOSM

In her study of the evolution of relationships among patients, doctors, society, and government, Dora Weiner uses the French Revolution and the Napoleonic era as a backdrop for exploring "the first modern attempt to grapple with the daunting problem of providing health care for an entire needy population."[1] Although not originally on the agenda of French revolutionaries, the "right" to health care quickly became linked to social programs designed to eliminate poverty. As noted by Dora B. Weiner, Professor and Charles E. Culpeper Scholar of the Medical Humanities at UCLA, such rights did not come without a high price tag: The national legislature in 1789 declared that it would compensate the blind, the deaf, and the mentally ill for their handicaps. People were called on to assume greater responsibility for their own health, for example, to protect public water sources and to avoid obvious health hazards. The notion of the citizen-patient emerged which promoted "the citizen-patients' partnership in the improvement of their health."[2]

Over the past two centuries, we have taken tremendous strides in our journey toward improved health care and wellness for the citizen-patient. Advances in diagnostic capabilities, in drug therapies and vaccinations, in clinical procedures and treatments, and in preventive medicine—to name but a few—have truly extended our range of capabilities. Now emboldened by information technology, we stretch even further toward that ideal of wellness for all, ever cognizant of the inspiration that the poet Robert Browning had in mind when he wrote, "A man's reach should extend his grasp, or what's a heaven for?"

The ideal is surely closer than we could have imagined five or ten years ago. Given the speed at which technologies change, we might modify that last statement to read five or ten *days* ago. But even more impressive than that speed is the significant shift in our thinking about "information" *per se,* rather than the technologies that generate, manage, and transmit it. Ken Murtha, executive director of Product and Customer Operations at Astra Merck, explains that distinction:

> In the past, it used to be that information was power. And the person who had that information was the dominant person in the relationship. Today, with information *technology,* you and I both share the same information—you, me, the customer, the supplier—at the exact same point of need. Therefore, all of us can have a better informed perspective on an outcome....This demands that you keep your own skills honed to a point where you still have believability. That's the challenge.[3]

A variation on the old adage that knowledge is power, Murtha's statement about information implies that we are not dealing only with "data and facts." Information technology is driving us toward the Health Care Infocosm, which, as we near the next century, brings social and cultural tumult as revolutionary as the great upheavals that began in Paris in 1789. Over the last few decades, we have watched the supply of information increase exponentially—so much so that today one often hears people talking about information overload. Mere possession of facts and data has never bestowed power. Throughout history—from the Greeks' defeat of the Persians at Salamis to NASA's astonishing photographs of Jupiter—we have won our greatest victories by skillfully and ingeniously applying our knowledge—not from simply owning raw data.

That said, it's important to acknowledge that information technology is *not* a panacea for all of modern society's health care troubles. It's not going to prevent epidemics, although it can give us advanced warning, guide our response, and keep the public informed about measures for guarding their own health. It's not going to equalize the quality of health care worldwide, although it will make it easier for people in remote regions to take advantage of health care innovations from any part of the world.

As we have emphasized throughout this book, information technology is destabilizing the health care industry at the same time that it enables enterprises to make the transformation to the Health Care Infocosm. That is today's undeniable "fact," and we must forge ahead to achieve the possible. Like all facts, that tells only part of the story.

Yes, the industry is in the midst of a great upheaval. Yes, that's traumatic. And yes, the landscape is hardly recognizable from moment to moment: Mergers, acquisitions, partnerships, and alliances are all part of the vast sea change now working its transformational magic. And the destabilization that information technology brings will span the career horizons of today's and tomorrow's leaders. Yet, we see a day in the not-so-distant future when the industry will settle down, and the enabling forces of information technology will show us better ways of doing business and delivering health care.

Here are the critical factors by which health care organizations need to measure their progress along the road to successful transformation to the Health Care Infocosm:

- Any health care enterprise that ignores market realities is at risk. Each organization must systematically transform itself to respond to shifts in customer demographics, needs, and expectations; changes in supplier networks; the appearance of new competitors; and the flight of members to rival organizations. In a true transformation, no component of an enterprise escapes scrutiny.

- An enterprise can aim for successful transformation only if its courageous leaders undertake an honest assessment of the organization's strengths and weaknesses. They must step away from their customary points of view to give consideration to whether they should base the transformation on current competencies or whether they should import new capabilities. To improve health care and wellness, the leaders must strive to develop cross-enterprise integration of processes and an understanding of information technology's power to achieve that integration.

- Virtual organizations will require new partnership and collaboration skills, fresh definitions of trust and cooperation, and a sense of what it means to share information of all sorts. Those are crucial ingredients in the realm of shared outcomes and the creation of knowledge, for which the industry as a whole will need to establish protective mechanisms and appropriate reward structures.

- Successful transformation relies more on foresight than on planning. It's easy to fantasize about the future; it's infinitely more difficult to ground vision in reality. Enabled and driven by information technology, such grounding requires a wide-ranging study of industry trends, past and current disruptions, and discontinuities that are likely to affect the future of the enterprise. What's more, all the weight of strategic planning does not equal an ounce of strategic intent: As Hamel and Prahalad point out, "strategic plans reveal more about today's problems than tomorrow's opportunities."[4]

- Meaningful change is not a onetime event. We need to consider it a process of striving toward the ideal rather than actually grasping it. Donald A. Schön, Ford Professor Emeritus and Senior Lecturer, School of Architecture and Planning at the Massachusetts Institute of Technology, reflects, "Change—its focus and energy—dissipates unless people stick with it. We make the mistake of thinking that there is an abstract organizational 'we' that makes change happen. That is not true. Change happens through the agency of persons in interaction with one another."[5] Accordingly, the industry's most courageous leaders must continue to inspire employees and customers with enthusiasm for a transformation that is both unavoidable and mutually beneficial. Occasionally that will promote the idea that the entire organization is balanced on a burning platform. Terror can be a great motivator, but perhaps a more effective approach calls for enlisting the organization's best people. They should engage in a team-directed effort to imagine the future enterprise and make it a reality.

- Information technology, no matter how powerful, can facilitate success only when people contribute their valuable knowledge and the benefit of their experience to the enterprise. Industry leaders must provide organizational space where change can take root and flourish. That creative space, however, is more than physical room: It embraces an attitude—an attitude that is uncommon in many of today's organizations—that risk-taking is essential and that mistakes, rather than threatening careers, open doors to new discoveries.

- Bold, decisive action is fundamental to an organization's success, and those leaders who proceed too cautiously—one small change here, another there—are likely to see their organizations swept aside by tidal waves of innovation both inside and outside the health care industry. Few enterprises will enjoy the luxury of forming an entirely new business on the order of the creation of Astra Merck, but organizations should emulate that mind-set by fostering the design of such greenfield enterprises.

- Perhaps the most critical of the success factors is the creation of a seamless health care fabric, fully integrating the work done by physicians, nurses, HMOs, insurance companies, pharmaceutical corporations, home health care services, government agencies, and the array of clinics and acute-care

> facilities worldwide. Transformed companies "do not focus unduly on the bottom line of profit. Instead, they concentrate primarily on customer performance measures, knowing that if they get this right the rest will follow."[6] To achieve sweeping solutions, health care enterprises must have customer service and patient-focused care firmly in place.[7]

Successful companies will have the most difficulty challenging the assumptions that made them rich. Still, if you consider the past, you'll recall how slow the giants were to adopt such revolutionary business processes as quick response and just-in-time inventory controls. Similarly, players in the airline industry had to realign their businesses in response to deregulation, computerized reservations systems, and, more recently, the on-line services that allow passengers to use their home computers to shop for the best deals.

The tumult in today's health care industry and the myriad changes that will confront us warn us to prepare for the unexpected. Some will fail to hear that call, but others will rub the sleep from their eyes and groggily begin their preparations. For some, the call will seem to be a tolling bell, or even a death knell. Others will rise to the roll of drums, preface to a grand work of promise. Yet others will hear the blast of a trumpet summoning them to begin the day's work. The real winners will be those who anticipate that clarion call: They will be meeting the challenges by signing up new members, making the most of the emerging information technologies, and smoothly responding to demands of customers, allied industries, and society at large. The Age of the Infocosm has already mounted its challenge.

Notes

Introduction

1. William Reinfeld. 1996. Internal communication to Gwendolyn B. Moore, February.

Chapter 1

1. Dan Wascoe. 1995. Mayo Clinic, Hussein Establish Satellite Link. *Star Tribune* (Minneapolis), November 17, n.p.
2. Steve Zurier. 1995. Telemedicine is Bringing Electronic Doctors Closer. *Government Computer News* 14.21, October 2: p. 56.
3. Dora B. Weiner. 1993. *The Citizen-Patient in Revolutionary and Imperial Paris.* Baltimore: Johns Hopkins University Press.
4. A Computer-Based Patient Record. 1996. *MD-Byline Information Services.* January.
5. David Bennahum. 1995. Docs for Docs. *Wired,* March, pp. 100–104.
6. Robert Fromberg. 1995. CPRI and the Future of Computer-based Patient Records. *Healthcare Financial Management* 49.7, July: p. 48.
7. Jack Sandman. 1995. Interview by Gwendolyn B. Moore. Columbus, Ohio. October 18.
8. Bill Richards. 1996. The 300-mile Stethoscope. *Wall Street Journal,* January 17, p. A1.
9. Alice LaPlante. 1995. A Virtual ER. *Forbes ASAP,* June 5, pp. 49–54.

10. Joyce E. Davis. 1995. Telemedicine Begins to Make its Case. *Fortune,* November 27, p. 44.
11. "HANC." 1996. Andersen Consulting memorandum.
12. Charles C. Mann. 1995. Guardian Angels. *Boston Magazine,* October, p. 39–42.
13. Jon Auerbach. 1995. The Doctor is On-line. *Boston Globe,* November 5, pp. 1, 28.
14. Bruce Knecht. 1996. Click! Doctor to Post Patient Files on Net. *Wall Street Journal,* February 20, pp. B1, 6.
15. Boon for Rural Areas—Computer Technology and High-speed Network Links are Changing the Face of Medical Services in Rural Communities. 1996. *Financial Times* (London), February 7, n.p.
16. Jay McWilliams. 1995. Telemedicine: A New Area of High-tech Specialization. *Pacific Business News,* December 18, n.p.
17. George C. Halvorson. 1993. *Strong Medicine.* New York: Random House.
18. Ibid.
19. Laura Johannes. 1996. Patients Delve into Databases to Second-guess Doctors. *Wall Street Journal,* February 21, pp. B1, 6.
20. Patient, Heal Thyself at Home. 1995. *Design Week,* September 1, p. 21.

CHAPTER 2

1. Peter H. Fuchs. 1995. Enterprise Transformation. Unpublished paper. Andersen Consulting, p. 23.
2. Andrew N. Morlet and Susan M. Oliver. 1995. Health Futures: Summary. Andersen Consulting.
3. Michael Hammer and James Champy. 1993. *Reengineering the Corporation: A Manifesto for Business Revolution.* New York: HarperCollins.
4. See Fuchs 1995.
5. Gary Hamel and C. K. Prahalad. 1989. Strategic Intent. *Harvard Business Review* 67.3, May–June: pp. 63–76.

6. Managed Care Market Overview: Phoenix, AZ. 1995. Boston: Charles J. Singer & Co., February.
7. Donald M. Berwick. 1995. Letter from Donald M. Berwick, M.D. The State of Health Care in America. *Business & Health* 13.3, Supplement C, pp. 4–5.
8. Janice Warila Young and Michael Martin. 1994. Information Management and the Evolving Delivery System. In *Health Care in America: An Industry in Transition.* Special issue of *The Marsh & McLennan Companies Quarterly* 23.4, Fall: pp. 43–51.
9. Warren R. Ross. 1994. Careers Still Flourish, but the Rules Have Changed. *Medical Marketing & Media* 29.12, December.
10. Andy Hines. 1994. Jobs and Infotech: Work in the Information Society. *The Futurist* 28.1, January, p. 9.

CHAPTER 3

1. Shawn Tully. 1995. America's Healthiest Companies. *Fortune,* June 12, pp. 98–106.
2. Future Focused: Payors as Information Providers. 1994. *Health Insurance & Technology,* April, n.p.
3. Peter Orton. 1994. Shared Care. *The Lancet* 344, no. 8934, November 19: p. 1413.
4. Erik Eckholm. 1994. While Congress Remains Silent, Health Care Transforms Itself. *New York Times,* December 18, pp. 1, 34.
5. Leroy L. Schwartz. 1994. The Medicalization of Social Problems: America's Special Health Care Dilemma. Washington: The AmHS Institute.
6. Gary Hamel and C. K. Prahalad. 1989. Strategic Intent. *Harvard Business Review* 67.3, May–June: pp. 63–76.
7. Ibid.
8. Ibid.

9. Peter H. Lewis. 1995. Microsoft Says It's Going after Internet Market. *New York Times,* December 8, pp. D1, 5.
10. Bart Ziegler. 1995. Internet Software Poses Big Threat to Notes, IBM's Stake in Lotus. *Wall Street Journal,* November 7, pp. A1, 5.

CHAPTER 4

1. Susan M. Oliver, Andrew N. Morlet, and Janeer Weerasingam. 1995. *Health Industry Practice, Asia Pacific.* Health Futures Forum (Australia). Andersen Consulting.
2. George Harrar. 1994. Pills 'n' Pads No More. *Forbes ASAP,* June 6, pp. 29–33.
3. Nancy Wysenski. 1995. Interview by John D. Rollins. Princeton, New Jersey. October 25.
4. The Health-Care Industry's Morning After. 1995. *The Economist,* July 15, pp. 45–46.
5. Richard A. Knox. 1996. Three Hospitals Form Cancer Partnership. *Boston Globe,* January 17, pp. 1, 4.
6. Ken Murtha. 1995. Interview by John D. Rollins. Princeton, New Jersey. October 25.
7. Donald A. Schön and Gwendolyn B. Moore. 1994. A Case Study of the Astra Merck Engagement: Extracting Lessons from Successes and Dilemmas. Internal Case Study. Andersen Consulting. September.
8. See Wysenski 1995.
9. Bill Gates. 1995. *The Road Ahead.* New York: Viking. [Excerpted in *Newsweek,* November 27, 1995, pp. 59–61, 64–68.]
10. Peering into 2010. 1994. *The Economist,* March 19, n.p.
11. Health Insurance Association. 1993. *Source book of health insurance data.* Washington, D.C.
12. Gwendolyn B. Moore. 1994. Lifestyles, Demographics and Epidemiology—Implications for the Future. Internal presentation. Andersen Consulting. December.

13. Jon Auerbach. 1995. The Doctor is On-line. *Boston Globe,* November 5, pp. 1, 28.
14. Gwendolyn B. Moore, Brian E. Reynolds, and David A. Rey. 1995. Information Architects: Data by Design. *HMO Magazine,* May–June, pp. 29–35.
15. Jane Linder and Gwendolyn B. Moore. 1991. Alliant Health System: A Vision of Total Quality. Harvard Business School, case no. 9–192–003, August 14.
16. See Moore, Reynolds, and Rey 1995.

CHAPTER 5

1. Toward the Virtual Enterprise. 1996. Andersen Consulting 1995 Annual Report.
2. Charles Roussel and Steven Kamman. 1995. Networking the Business System: Why Alliances are Critical to Building the Networked Corporation of the Twenty-first Century. Unpublished manuscript. Andersen Consulting.
3. Shoshana Zuboff. 1988. *In the Age of the Smart Machine: The Future of Work and Power.* New York: Basic Books, p. 222.
4. Clifford S. Hakim. 1994. You're the Boss Now. *National Business Employment Weekly,* September 11–17, p. 1.
5. The Future of European Health Care. 1993. Andersen Consulting in cooperation with Burson-Marsteller.
6. David R. Olmos. 1996. Hospitals Reinvent Themselves. *Los Angeles Times,* January 11, p. A1.
7. See Zuboff 1988: pp. 286, 287.
8. Charles Handy. 1990. *The Age of Unreason.* Boston: Harvard Business School Press, p. 141.
9. Ibid, p. 142.
10. Michael Schrage. 1995. *No More Teams! Mastering the Dynamics of Creative Collaboration.* New York: Doubleday, p. 32.

11. Victoria Griffiths. 1996. Cyber Doctors: Patients and Medical Staff Have Taken to the Internet and On-line Services. *Financial Times,* January 12, p. 10.
12. See Schrage 1995.
13. Scott A. Snell and James W. Dean, Jr. 1994. Strategic Compensation for Integrated Manufacturing: The Moderating Effects of Jobs and Organizational Inertia. *Academy of Management Journal* 37.5: pp. 1109–1140.
14. Michael A. Verespej. 1994. New Responsibilities? New Pay! *Industry Week,* August 15, pp. 11, 14, 16, 18, 22.
15. J. Duncan Moore, Jr. 1995. Raises Slow to a Trickle: Compensation Survey's Results Reflect Healthcare's Uncertainties. *Modern Healthcare,* June 26, pp. 47–52.
16. See Snell and Dean 1994: p. 1118.
17. Ziva Freiman. 1994. Hype vs. Reality: The Changing Workplace. *Progressive Architecture* 75.3: pp. 48.
18. Barry K. Spiker, David A. Miron, Eric L. Lesser, and David H. Jackson. 1994. Managing Change in the Health Care Industry. In *Health Care in America: An Industry in Transition.* Special issue of *The Marsh & McLennan Companies Quarterly* 23.4, Fall: pp. 53–61.
19. Michael D. McDonald. 1994. Telecognition for Improving Health. *Healthcare Forum Journal,* March–April, pp. 18, 20.
20. Ibid.
21. Ibid.
22. Ibid.
23. Sabra Chartrand. 1995. Why is This Surgeon Suing? *New York Times,* June 8, pp. D1, 5.

CHAPTER 7

1. Dora B. Weiner. 1993. *The Citizen-Patient in Revolutionary and Imperial Paris.* Baltimore: Johns Hopkins University Press. p. xv.
2. Ibid, pp. 3–14.

3. Ken Murtha. 1995. Interview by John D. Rollins. Princeton, New Jersey. October 25.
4. Gary Hamel and C. K. Prahalad. 1989. Strategic Intent. *Harvard Business Review* 67.3, May–June, pp. 63–76.
5. Donald A Schön. 1996. Interview by Gwendolyn B. Moore. Boston, Massachusetts. February 17.
6. Nicholas Edwards. 1994. Transforming Marketing for a Profitable Future. *Scrip Magazine,* June, p. 28.
7. Ned Troup and Steve Rushing. 1992. Working Smarter with Patient-Focused Care. *Southern Hospitals,* July–August, pp. 13–14, 27.